SpringerBriefs in Applied Sciences and Technology

Computational Mechanics

For further volumes:
http://www.springer.com/series/8886

Marcelo J. S. de Lemos

Turbulent Impinging Jets into Porous Materials

 Springer

Marcelo J. S. de Lemos
Departamento de Energia—IEME
Instituto Tecnólogico de Aeronáutica
São José dos Campos
12228-900
Brazil

ISSN 2191-5342   ISSN 2191-5350 (electronic)
ISBN 978-3-642-28275-1   ISBN 978-3-642-28276-8 (eBook)
DOI 10.1007/978-3-642-28276-8
Springer Heidelberg New York Dordrecht London

Library of Congress Control Number: 2012932405

Printed on acid-free paper

Springer is part of Springer Science+Business Media (www.springer.com)

"Ars longa, vita brevis, occasio praeceps, experimentum periculosum, iudicium difficile" (art is long, life is short, opportunity fleeting, experiment treacherous, judgment difficult)

Hippocrates (c. 460 BC–c. 370 BC)

To my mother
Maria da Glória Santos de Lemos

Preface

For more than two decades now, at the Instituto Tecnológico de Aeronáutica (ITA), Brazil, I have taught courses on thermal sciences. In the middle of the 1990s, we got interested in the subject of modeling flows through permeable structures. Besides the most promising realm of applications for such studies, namely Oil and Gas industries, other promising areas such as Energy, Aeronautics, Aerospace and Defense, became also target for new applied technology to be developed based on the fundamentals detailed in several journal articles and books published on this subject by our group.

Among many applications that we foresaw in years back, was the possibility of enhancing or damping heat transfer rates from surfaces subjected to heating or cooling by an impinging jet. Then, a series of systematic studies were carried out describing the advantages, or otherwise, of having a solid porous matrix attached to a surface that is hit by a fluid. This book intends to present such studies in a self contained and organized way.

I am much thankful to and would like to express my appreciation to Professor Dr.-Ing. Andreas Öchsner, Editor-in-Chief of the Springer book series on "Advanced Structured Materials", among other editorships, and Dr. Christoph Baumann, Springer Senior Engineering Editor, for their encouragement in putting forward this project.

The continuous support from our research funding agencies in Brazil, namely, CAPES, CNPq and FAPESP, are greatly appreciated. Among all my former students who have contributed to our overall research goals in the last fifteen years, special thanks are due to master graduates D.R. Graminho, C. Fischer and F.T. Dórea for their efforts on the research topic here discussed. Finally, I am thankful to Mrs. Stefani Montemagni for her careful and skillful typing of the manuscript.

São José dos Campos, December 2011   Marcelo J.S. de Lemos

Overview

This book presents numerical results for a turbulent jet impinging against a flat plane covered with a layer of permeable and thermally conducting material. The mathematical model describing the mean and statistical fields take into consideration the heterogeneity of the porous medium, formed by a light-weight solid structure containing voids randomly distributed, and a working fluid that crosses such permeable structure. The voids are assumed to be of sufficient size for turbulence to exist, or say, the fluid undergoes transitions and, at the pore level, it is described as a fully turbulent flow.

When an impermeable wall is covered with a layer of porous material, the computational domain may consider not only the porous structure itself, but also an additional void space over the layer. There, in this homogeneous space, the working fluid flows freely and throughout this text it is referred to as a "clear" region. Also, inside the porous material distinct energy equations are considered for the porous layer attached to the wall and for the fluid that permeates it.

Parameters such as Reynolds number, porosity, permeability, thickness, and thermal conductivity of the porous layer are varied in order to analyze their effects on the local distribution of Nu. The macroscopic equations for mass, momentum, and energy are obtained based on volume-average concept. The numerical technique employed for discretizing the governing equations was the control volume method with a boundary-fitted nonorthogonal coordinate system. The SIMPLE algorithm was used to handle the pressure-velocity coupling. Results indicate that inclusion of a porous layer eliminates the peak in Nu at the stagnation region. For highly porous and highly permeable material, simulations indicate that the integral heat flux from the wall is enhanced when a thermally conducting porous material is attached to the surfaced.

Contents

1	**Introduction**	1
	1.1 Configurations of Impinging Jets	2
	References	5
2	**Mathematical Modeling of Turbulence in Porous Media**	7
	2.1 Local Instantaneous Transport Equations	7
	2.2 Double-Decomposition of Variables	8
	2.3 Macroscopic Flow Equations	8
	2.4 One-Energy Equation Model	10
	2.5 Two-Energy Equation Model	13
	2.5.1 Turbulence	15
	2.5.2 Interfacial Heat Transfer, h_i	16
	2.5.3 Non-Dimensional Parameters	16
	2.6 Boundary Conditions and Numerical Details	17
	2.6.1 Flow Boundary and Interface Conditions	17
	2.6.2 Heat Boundary and Interface Conditions	17
	2.6.3 Numerical Details	18
	References	18
3	**Flow Structure of Impinging Jets**	21
	3.1 Clear Chamber	22
	3.1.1 Mean Flow	22
	3.1.2 Turbulent Field	26
	3.2 Porous Medium	27
	3.2.1 Mean Flow	27
	3.2.2 Turbulent Field	33
	References	35

4 Heat Transfer Using the Local Thermal Equilibrium Model   37
  4.1   Input Parameters for the LTE Model   37
  4.2   Grid Independence Studies for the LTE Model   38
  4.3   Clear Channel   38
  4.4   Channel with Porous Layer   41
        4.4.1   Effect of Porosity, ϕ   41
        4.4.2   Effect of Channel Blockage, h/H   44
        4.4.3   Effect of Darcy Number, Da   46
        4.4.4   Effect of Thermal Conductivity Ration k_s/k_f   49
  4.5   Integral Wall Heat Flux   50
  References   53

**5 Heat Transfer Using the Local Thermal
  Non-Equilibrium Model**   55
  5.1   Input Parameters for the LTNE Model   55
  5.2   Grid Independence Studies for the LTNE Model   55
  5.3   Empty Channel   56
  5.4   Channel with Porous Layer   58
        5.4.1   Effect of Reynolds, Re   58
        5.4.2   Effect of Porosity, ϕ   60
        5.4.3   Effect of Channel Blockage, h/H   62
        5.4.4   Effect of Darcy Number, Da   64
        5.4.5   Effect of Solid-to-Fluid Thermal
                Conductivity Ratio k_s/k_f   67
  5.5   Integral Wall Heat Flux   69
  References   72

6 Concluding Remarks and Future Work   75
  References   75

Nomenclature

A_i	Interfacial area between solid and fluid in a porous medium
B	Jet width
c_F	Forchheimer coefficient
Da	Darcy number, $Da = K/H^2$
H	Channel height
h	Porous layer thickness, film coefficient
K	Permeability
k_{eff}	Effective thermal conductivity
k	(1) Thermal conductivity; (2) Turbulence kinetic energy per mass unit, $k = \overline{u' \cdot u'}/2$
L	Channel length
Nu	Nusselt number
p	Thermodynamic pressure
q_w	Integral wall heat flux
q_w^ϕ	Integral wall heat flux will porous layer
$\langle p \rangle^i$	Intrinsic (fluid) average of pressure p
Re	Reynolds number based on the jet width, $Re = \rho v_0 B / \mu$
S_φ	Source term
T	Temperature
$\langle \mathbf{u} \rangle^i$	Intrinsic (fluid) average of $\mathbf{u}$
$\mathbf{u}_D$	Darcy velocity vector (volume average over $\mathbf{u}$) $= \phi \langle \mathbf{u} \rangle^i$
x, y	Cartesian coordinates
ϕ	Porosity

Greek Symbols

μ Dynamic viscosity
ρ Density
v Kinematic viscosity
σ_{t_ϕ} Macroscopic turbulent Prandtl number
ϕ Related to porous medium

Symbols, Subscripts, Special Characters

s, f solid, fluid
w wall
0 Inlet conditions

Chapter 1
Introduction

Impinging jets are often used in industrial applications for enhancing or damping localized heat transfer rates. When the flow is turbulent, thin boundary layers are located inside the stagnation zone, promoting even further cooling, heating or drying processes. Applications of such systems include metals cooling, glass tempering, electronics cooling, drying of textiles products and paper, to mention a few. In this book, two flow configurations are investigated, namely axisymmetric confined arrangements and two-dimensional planar jets. A fluid jet enters a cylindrical chamber through an aperture in an upper disk. An annular clearance between the cylinder lateral wall and the disc allows fluid to flow out of the enclosure.

The majority of results in the open literature are mostly related to impinging jets under high mass flow conditions that reaches a bare surface. Studies considering two-dimensional impinging jets with low Reynolds number, also onto uncovered walls, are presented in Gardon and Akfirat [1], who experimentally obtained local and averaged heat transfer coefficients. Sparrow and Wong [2] made use of the well-known heat and mass transfer analogy and took experimental data on local mass transfer for a two-dimensional impinging jet. Results were then converted to heat transfer using the mentioned technique. Chen et al. [3] experimentally and numerically analyzed mass and heat transfer induced by a two-dimensional laminar jet. Chiriac and Ortega [4] performed numerical simulations in steady and transitory regime for a two-dimensional jet impinging against a plate with constant temperature. Additional works on impinging jets on bare surfaces for oscillating [5] and steady-state regimes are also found in the literature [6, 7].

In addition, in the recent years a number of research papers covered a wide range of studies in porous media [8–11], including flows parallel to a layer of porous material [12] and across permeable baffles [13, 14] and porous inserts [15]. Investigation on configurations concerning perpendicular jets into a porous core is much needed for optimization of heat sinks attached to solid surfaces. However, studies of porous medium under impinging jets are, unfortunately, yet very scarce in the literature. An example found are those given by numerical simulations of Kim and Kuznetsov [16], who investigated optimal characteristics of impinging

M. J. S. de Lemos, *Turbulent Impinging Jets into Porous Materials*,
SpringerBriefs in Computational Mechanics,
DOI: 10.1007/978-3-642-28276-8_1, © The Author(s) 2012

jets into heat sinks. Other innovative applications of impinging jets, such as fiber hydroentanglement, can also be found in the recent literature [17, 18].

Examples of work on impinging jets are presented by Prakash et al. [19, 20] who obtained a flow visualization of turbulent jets impinging against a porous medium. Also, Fu and Huang [21] evaluated the thermal performance of different porous layers under a impinging jet and Jeng and Tzeng [22] studied the hydro-dynamic and thermal performance of a jet impinging on a metallic foam. Recently, the flow structure of a jet impinging on a porous layer has been investigated for both laminar [23] and turbulent [24] regimes. In Ref. [24], the turbulence model described in detail by [25] was applied. Earlier, Rocamora and de Lemos [26] had added thermal modeling for the treatment of a permeable medium considering thermal equilibrium between the solid and the fluid. Later, the work on isothermal impinging jets was extended to involve laminar [27] as well as turbulent [28] thermal analyses, being both such works based on the local thermal equilibrium hypothesis (LTE) [26]. Subsequently, the one-energy equation model of [26] was extended to account for energy transport in each phase, using, for achieving such goal, the so-called local thermal-non-equilibrium model (LTNE). The LTNE closure was then applied to both flow regimes, namely laminar [29] and turbulent [30] flows. More recently, the local non-thermal equilibrium model was applied to simulate both a laminar impinging jet into a porous layer [31] as well as turbulent flow regime onto the same geometry [32].

This book will review and compiles results in Ref. [24] for the structure of the turbulent flow of jets in addition to simulations using the one-energy equation model [28] as well as applications of the two-energy equation approach [32]. By that, one can exploit the advantages, if any, in adding a layer of porous material to enhance the overall heat transferred from a surface. The objective is to evaluate the use of different thermal models in predicting impinging jets into a porous material. Comparisons using both LTE and LTNE models are presented. In the text to follow, the LTE model is also given the acronym 1EEM, for "One-Energy Equation Model", whereas the LTNE approach is recalled as 2EEM, standing for "Two-Energy Equation Model".

1.1 Configurations of Impinging Jets

The first problem considered is schematically presented in Fig. 1.1a, which shows the same geometry used by Prakash et al. [19, 20]. The reason for choosing the same geometry as in the pioneering work of [19, 20] was to be able to validate the computations herein with the experimental data provided by them. As shown in the Fig. 1.1b, a fluid jet enters a cylindrical chamber through an aperture in an upper disk. An annular clearance between the cylinder lateral wall and the disc allows fluid to flow out of the enclosure. The incoming jet diameter, D_j, is 0.019 m and the inner cylinder diameter, D, is 0.39 m. The clearance between the cylinder and

Fig. 1.1 Axis-symmetric
flows: **a** confined jet
impinging against a porous
layer, **b** cross section view
and nomenclature

(a)

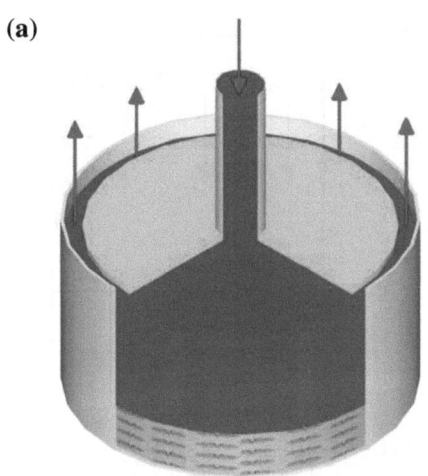

(b)

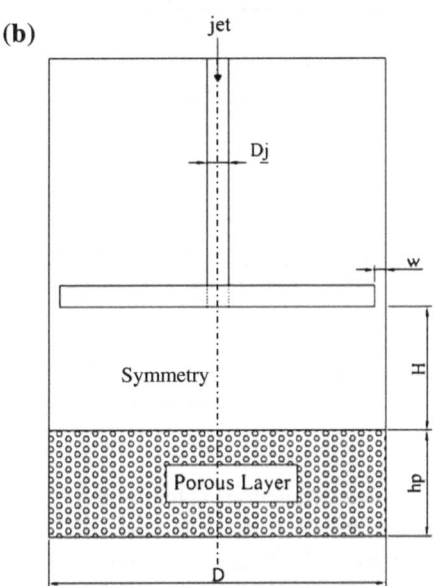

the disc holding the jet has a width, w, equal to 0.005 m. At the bottom of the
chamber, a layer of porous material covers the surface and is hit by the incoming
jet. Three different heights of fluid column above the porous substrate, H, namely
0.05, 0.1 and 0.15 m, were used in the simulations. Two thicknesses of the porous
layer, h_p, were considered, namely 0.05 and 0.1 m. The average velocity of the
incoming jet was 1.6 m/s representing a Reynolds number of 30,000, which was
based on the jet exit diameter. Different velocities were also used when evaluating
the influence of Reynolds number on the main flow. The two other values 1.0 and
2.5 m/s corresponded to $Re = 18,900$ and 47,000, respectively. Turbulent results

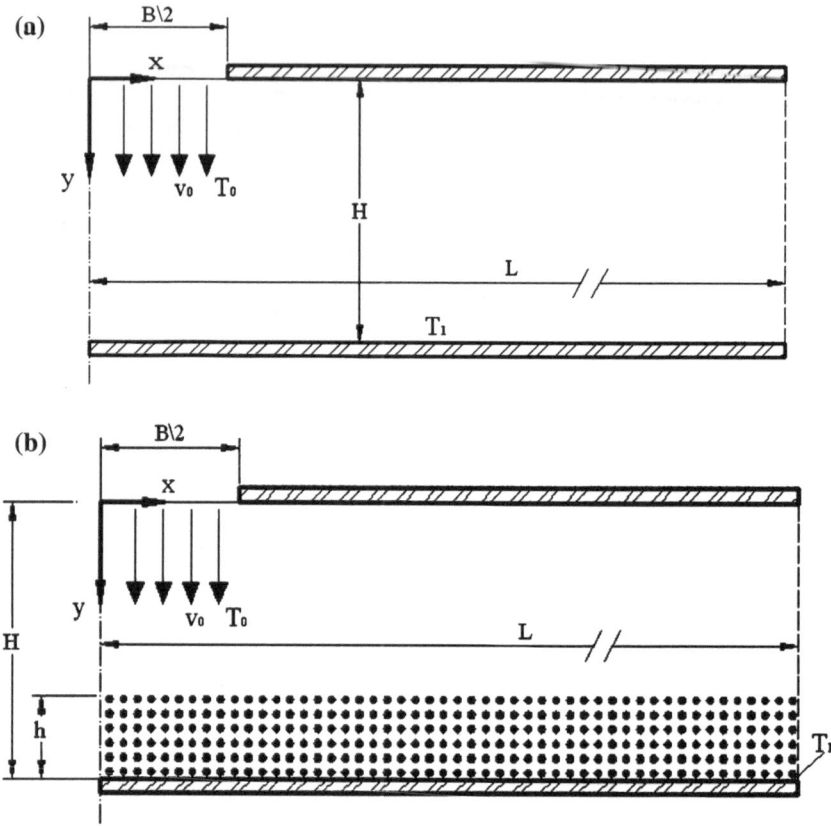

Fig. 1.2 Two-dimensional planar flows: **a** confined impinging jet on a flat plate—clear medium case, **b** confined impinging jet on a plate covered with a layer of porous material—porous case

for stream function, velocity and turbulence kinetic energy profiles are compared here with those from Prakash et al. [19, 20]. Also considered here are the two-dimensional planar cases detailed in Fig. 1.2a. A turbulent jet with uniform velocity v_o and constant temperature T_o enters through a gap into a channel with height H and length $2L$. Fluid impinges normally against the bottom plate yielding a two-dimensional confined impinging jet configuration. The width of the inlet nozzle is B and the bottom plate temperature, T_1, is maintained constant and 38.5 K above the temperature of the incoming jet, T_o. In a different configuration, the bottom surface is covered with a porous layer of height h (Fig. 1.2b). In both cases, the flow is assumed to be two-dimensional, turbulent, incompressible and steady. Also here the porous medium is taken to be homogeneous, rigid and inert. Fluid properties are constant and gravity effects are neglected.

The boundary conditions for the problem in Fig. 1.2 are: (a) constant velocity and temperature profiles of the entering jet, (b) no slip condition on the walls, (c) symmetry condition in $x = 0$, (d) fully developed flow at channel exit ($x = L$). At the bottom plate ($y = H$), constant temperature condition is assumed whereas along the upper wall, for $B/2 < x \leq L$, null heat flux condition prevails.

References

1. R. Gardon, J.C. Akfirat, Heat transfer characteristics of impinging two-dimensional air jets. J. Heat Transf. **101**, 101–108 (1966)
2. E.M. Sparrow, T.C. Wong, Impinging transfer coefficients due to initially laminar slot jets. Int. J. Heat Mass. Transf. **18**, 597–605 (1975)
3. M. Chen, R. Chalupa, A.C. West, V. Modi, High schmidt mass transfer in a laminar impinging slot jet. Int. J. Heat Mass. Transf. **43**, 3907–3915 (2000)
4. V.A. Chiriac, A. Ortega, A numerical study of the unsteady flow and heat transfer in a transitional confined slot jet impinging on an isothermal surface. Int. J. Heat Mass. Transf. **45**, 1237–1248 (2002)
5. T. Dermircan, H. Turkoglu, The numerical analysis of oscillating impinging jets. Numer. Heat Transf. Part A Appl. **58**(2), 146–161 (2010)
6. N. Uddin, S.O. Neumann, B. Weigand, B.A. Younis, Large-eddy simulations and heat-flux modeling in a turbulent impinging jet. Numer. Heat Transf. Part A Appl. **55**(10), 906–930 (2009)
7. M.K. Isman, E. Pulat, A.B. Etemoglu, M. Can, Numerical investigation of turbulent impinging jet cooling of a constant heat flux surface. Numer. Heat Transf. Part A Appl. **53**(10), 1109–1132 (2008)
8. Y. Zhang, X.F. Peng, I. Conte, Heat and mass transfer with condensation in non-saturated porous media. Numer. Heat Transf. Part A Appl. **52**, 1081–1100 (2007)
9. M.E. Taskin, A.G. Dixon, E.H. Stitt, CFD study of fluid flow and heat transfer in a fixed bed of cylinders. Numer. Heat Transf. Part A Appl. **52**(3), 203–218 (2007)
10. X.H. Wang, M. Quintard, G. Darche, Adaptive mesh refinement for one-dimensional three-phase flow with phase change in porous media. Numer. Heat Transf. Part A Appl. **50**(4), 315–352 (2006)
11. A. Mansour, A. Amahmid, M. Hasnaoui, M. Bourich, Multiplicity of solutions induced by thermosolutal convection in a square porous cavity heated from below and submitted to horizontal concentration gradient in the presence of Soret effect. Numer. Heat Transf. Part A Appl. **49**(1), 69–94 (2006)
12. A.V. Kuznetsov, L. Cheng, M. Xiong, Effects of thermal dispersion and turbulence in forced convection in a composite parallel-plate channel: investigation of constant wall heat flux and constant wall temperature cases. Numer. Heat Transf. Part A Appl. **42**(4), 365–383 (2002)
13. B.M.D. Miranda, N.K. Anand, Convective heat transfer in a channel with porous baffles. Numer. Heat Transf. Part A Appl. **46**(5), 425–452 (2004)
14. N.B. Santos, M.J.S. de Lemos, Flow and heat transfer in a parallel-plate channel with porous and solid baffles. Numer. Heat Transf. Part A Appl. **49**(5), 471–494 (2006)
15. M. Assato, M.H.J. Pedras, M.J.S. de Lemos, Numerical solution of turbulent channel flow past a backward-facing-step with a porous insert using linear and non-linear k-ε models. J. Porous Media **8**(1), 13–29 (2005)
16. S.Y. Kim, A.V. Kuznetsov, Optimization of pin-fin heat sinks using anisotropic local thermal nonequilibrium porous model in a jet impinging channel. Numer. Heat Transf. Part A Appl. **44**(8), 771–787 (2003)
17. P. Xiang, A.V. Kunetsov, A.M. Seyam, A porous medium model of the hydroentanglement process. J. Porous Media **11**, 35–49 (2008)

18. P, Xiang, A.V. Kunetsov, Simulation of shape dynamics of a long flexible fiber in a turbulent flow in the hydroentanglement process. Int. Commun. Heat Mass Transf. **35**, 529–534 (2008)

19. M. Prakash, F.O. Turan, Y. Li, J. Manhoney, G.R. Thorpe, Impinging round jet studies in a cylindrical enclosure with and without a porous layer: Part I: Flow visualizations and simulations. Chem. Eng. Sci. **56**, 3855–3878 (2001)

20. M. Prakash, F.O. Turan, Y. Li, J. Manhoney, G.R. Thorpe, Impinging round jet studies in a cylindrical enclosure with and without a porous layer: Part II: DLV measurements and simulations. Chem. Eng. Sci. **56**, 3879–3892 (2001)

21. W.-S. Fu, H.-C. Huang, Thermal performance of different shape porous blocks under an impinging jet. Int. J. Heat Mass Transf. **40**(10), 2261–2272 (1997)

22. T.-Z. Jeng, S.-C. Tzeng, Numerical study of confined slot jet impinging on porous metallic foam heat sink. Int. J. Heat Mass Transf. **48**, 4685–4694 (2005)

23. D.R. Graminho, M.J.S. de Lemos, Laminar confined impinging jet into a porous layer. Numer. Heat Transf. Part A Appl. **54**(2), 151–177 (2008)

24. D.R. Graminho, M.J.S. de Lemos, Simulation of turbulent impinging jet into a cylindrical chamber with and without a porous layer at the bottom. Int. J. Heat Mass Transf. **52**, 680–693 (2009)

25. M.J.S. de Lemos, *Turbulence in Porous Media: Modeling and Applications* (Elsevier, Amsterdam, 2006)

26. F.D. Rocamora Jr, M.J.S. de Lemos, Analysis of convective heat transfer of turbulent flow in saturated porous media. Int. Commun. Heat Mass Transf. **27**(6), 825–834 (2000)

27. M.J.S. de Lemos, C. Fischer, Thermal analysis of an impinging jet on a plate with and without a porous layer. Numer. Heat Transf. Part A Appl. **54**, 1022–1041 (2008)

28. C. Fischer, M.J.S. de Lemos, A turbulent impinging jet on a plate covered with a porous layer. Numer. Heat Transf. Part A Appl. **58**, 429–456 (2010)

29. M.B. Saito, M.J.S. de Lemos, Interfacial heat transfer coefficient for non-equilibrium convective transport in porous media. Int. Commun. Heat Mass Transf. **32**(5), 666–676 (2005)

30. M.B. Saito, M.J.S. de Lemos, A macroscopic two-energy equation model for turbulent flow and heat transfer in highly porous media. Int. J. Heat Mass Transf. **53**(11–12), 2424–2433 (2010)

31. F.T. Dórea, M.J.S. de Lemos, Simulation of laminar impinging jet on a porous medium with a thermal non-equilibrium model. Int. J. Heat Mass Transf. **53**, 5089–5101 (2010)

32. M.J.S. de Lemos, F.T. Dórea, Simulation of turbulent impiging jet into a layer of porous material using a two-energy equation model. Numer. Heat Transf. Part A Appl. **59**(10), 769–798 (2011)

Chapter 2
Mathematical Modeling of Turbulence in Porous Media

As mentioned, the flow model here employed is described in Graminho and de Lemos [1] whereas thermal modeling is detailed in [2], including now the energy equation for heat transfer calculations. Therein, details can be found. As most of the theoretical development is readily available in the open literature, the governing equations will be just presented and details about their derivations can be obtained in the mentioned papers. Essentially, local instantaneous equations are volume-averaged using appropriate mathematical tools [3].

At the jet exit, a fully developed profile for velocity, k and ε was imposed. At the flow outlet through the clearance of width w, a zero diffusion flux condition was set. On the walls, a non-slip condition was applied and at the centerline of the cylinder, the symmetry condition was used.

2.1 Local Instantaneous Transport Equations

The governing equations for the flow and energy for an incompressible fluid are given by:

$$\text{Continuity:} \ \nabla \cdot \mathbf{u} = 0. \tag{2.1}$$

$$\text{Momentum:} \ \rho \left[\frac{\partial \mathbf{u}}{\partial t} + \nabla \cdot (\mathbf{u}\mathbf{u}) \right] = -\nabla p + \mu \nabla^2 \mathbf{u}. \tag{2.2}$$

$$\text{Energy-Fluid Phase:} \ (\rho c_p)_f \left\{ \frac{\partial T_f}{\partial t} + \nabla \cdot (\mathbf{u} T_f) \right\} = \nabla \cdot (k_f \nabla T_f) + S_f. \tag{2.3}$$

$$\text{Energy-Solid Phase (Porous Matrix)}: \ (\rho c_p)_s \frac{\partial T_s}{\partial t} = \nabla \cdot (k_s \nabla T_s) + S_s. \tag{2.4}$$

where the subscripts f and s refer to fluid and solid phases, respectively. Here, T is the temperature, k_f and k_s are the fluid and solid thermal conductivities,

M. J. S. de Lemos, *Turbulent Impinging Jets into Porous Materials*,
SpringerBriefs in Computational Mechanics,
DOI: 10.1007/978-3-642-28276-8_2, © The Author(s) 2012

respectively, c_p is the specific heat and S is the heat generation term. If there is no heat generation either in the solid or in the fluid, one has $S_f = S_s = 0$.

2.2 Double-Decomposition of Variables

Macroscopic transport equations for turbulent flow in a porous medium are obtained through the simultaneous application of time and volume average operators over a generic fluid property φ. Such concepts are defined as.

$$\bar{\varphi} = \frac{1}{\Delta t} \int_t^{t+\Delta t} \varphi \, dt, \text{ with } \varphi = \bar{\varphi} + \varphi' \tag{2.5}$$

$$\langle \varphi \rangle^i = \frac{1}{\Delta V_f} \int_{\Delta V_f} \varphi \, dV; \langle \varphi \rangle^v = \phi \langle \varphi \rangle^i; \phi = \frac{\Delta V_f}{\Delta V}, \text{ with } \varphi = \langle \varphi \rangle^i + {}^i\varphi \tag{2.6}$$

where ΔV_f is the volume of the fluid contained in a Representative Elementary Volume (REV) ΔV, intrinsic average and volume average are represented, respectively, by $\langle \ \rangle^i$ and $\langle \ \rangle^v$. The double decomposition idea fully described in [4], combines Eqs. (2.5) and (2.6) and can be summarized as:

$$\overline{\langle \varphi \rangle^i} = \langle \bar{\varphi} \rangle^i; {}^i\bar{\varphi} = \overline{{}^i\varphi}; \overline{\langle \varphi' \rangle^i} = \langle \varphi \rangle^{i'} \tag{2.7}$$

and,

$$\left. \begin{array}{l} \varphi' = \langle \varphi' \rangle^i + {}^i\varphi' \\ {}^i\varphi = \overline{{}^i\varphi} + {}^i\varphi' \end{array} \right\} \quad \text{where} \quad {}^i\varphi' = \varphi' - \langle \varphi' \rangle^i = {}^i\varphi - \overline{{}^i\varphi} \tag{2.8}$$

Therefore, the quantity φ can be expressed by either,

$$\varphi = \overline{\langle \varphi \rangle^i} + \langle \varphi \rangle^{i'} + \overline{{}^i\varphi} + {}^i\varphi' \tag{2.9}$$

or

$$\varphi = \langle \bar{\varphi} \rangle^i + {}^i\bar{\varphi} + \langle \varphi' \rangle^i + {}^i\varphi'. \tag{2.10}$$

The term ${}^i\varphi'$ can be viewed as either the temporal fluctuation of the spatial deviation or the spatial deviation of the temporal fluctuation of the quantity φ.

2.3 Macroscopic Flow Equations

When the average operators (2.5) and (2.6) are simultaneously applied over Eqs. (2.1) and (2.2), macroscopic equations for turbulent flow are obtained. Volume integration is performed over a Representative Elementary Volume (REV) [3], resulting in,

$$\text{Continuity: } \nabla \cdot \bar{\mathbf{u}}_D = 0. \tag{2.11}$$

where, $\bar{\mathbf{u}}_D = \phi\langle\bar{\mathbf{u}}\rangle^i$ and $\langle\bar{\mathbf{u}}\rangle^i$ identifies the intrinsic (liquid) average of the time-averaged velocity vector $\bar{\mathbf{u}}$.

Momentum:

$$\rho\left[\frac{\partial\bar{\mathbf{u}}_D}{\partial t} + \nabla\cdot\left(\frac{\bar{\mathbf{u}}_D\bar{\mathbf{u}}_D}{\phi}\right)\right] = -\nabla(\phi\langle\bar{p}\rangle^i) + \mu\nabla^2\bar{\mathbf{u}}_D - \nabla\cdot(\rho\phi\langle\overline{\mathbf{u}'\mathbf{u}'}\rangle^i)$$
$$- \left[\frac{\mu\phi}{K}\bar{\mathbf{u}}_D + \frac{c_F\phi\rho|\bar{\mathbf{u}}_D|\bar{\mathbf{u}}_D}{\sqrt{K}}\right] \tag{2.12}$$

where the last two terms in Eq. (2.12) represent the Darcy and Forchheimer or form drags. The symbol K is the porous medium permeability, c_F is the form drag or Forchheimer coefficient, $\langle\bar{p}\rangle^i$ is the intrinsic average pressure of the fluid and ϕ is the porosity of the porous medium.

The macroscopic Reynolds stress, $-\rho\phi\langle\overline{\mathbf{u}'\mathbf{u}'}\rangle^i$, appearing in Eq. (2.12) is given as,

$$-\rho\phi\langle\overline{\mathbf{u}'\mathbf{u}'}\rangle^i = \mu_{t_\phi}2\langle\bar{\mathbf{D}}\rangle^v - \frac{2}{3}\phi\rho\langle k\rangle^i\mathbf{I} \tag{2.13}$$

where,

$$\langle\bar{\mathbf{D}}\rangle^v = \frac{1}{2}\left[\nabla(\phi\langle\bar{\mathbf{u}}\rangle^i) + [\nabla(\phi\langle\bar{\mathbf{u}}\rangle^i)]^T\right] \tag{2.14}$$

is the macroscopic deformation tensor, $\langle k\rangle^i = \langle\overline{\mathbf{u}'\cdot\mathbf{u}'}\rangle^i/2$ is the intrinsic turbulent kinetic energy, and μ_{t_ϕ}, is the turbulent viscosity, which is modeled similarly to the case of clear flow, in the form,

$$\mu_{t_\phi} = \rho f_\mu c_\mu \frac{\langle k\rangle^{i^2}}{\langle\varepsilon\rangle^i} \tag{2.15}$$

The intrinsic turbulent kinetic energy per unit mass and its dissipation rate are governed by the following equations,

$$\rho\left[\frac{\partial}{\partial t}(\phi\langle k\rangle^i) + \nabla\cdot(\bar{\mathbf{u}}_D\langle k\rangle^i)\right] = \nabla\cdot\left[\left(\mu + \frac{\mu_{t_\phi}}{\sigma_k}\right)\nabla(\phi\langle k\rangle^i)\right] - \rho\langle\overline{\mathbf{u}'\mathbf{u}'}\rangle^i : \nabla\bar{\mathbf{u}}_D$$
$$+ c_k\rho\frac{\phi\langle k\rangle^i|\bar{\mathbf{u}}_D|}{\sqrt{K}} - \rho\phi\langle\varepsilon\rangle^i \tag{2.16}$$

$$\rho\left[\frac{\partial}{\partial t}(\phi\langle\varepsilon\rangle^i) + \nabla\cdot(\bar{\mathbf{u}}_D\langle\varepsilon\rangle^i)\right] = \nabla\cdot\left[\left(\mu + \frac{\mu_{t_\phi}}{\sigma_\varepsilon}\right)\nabla(\phi\langle\varepsilon\rangle^i)\right]$$
$$+ c_1(-\rho\langle\overline{\mathbf{u}'\mathbf{u}'}\rangle^i : \nabla\bar{\mathbf{u}}_D)\frac{\langle\varepsilon\rangle^i}{\langle k\rangle^i}$$
$$+ c_2c_k\rho\frac{\phi\langle\varepsilon\rangle^i|\bar{\mathbf{u}}_D|}{\sqrt{K}} - c_2f_\mu\rho\phi\frac{\langle\varepsilon\rangle^{i^2}}{\langle k\rangle^i} \tag{2.17}$$

where, $\sigma_k = 1.4$, $\sigma_\varepsilon = 1.3$, $c_1 = 1.50$, $c_2 = 1.90$, $c_\mu = 0.09$ and $c_k = 0.28$ are non-dimensional constants tuned for the Low-Reynolds number k-ε model whereas f_2 and f_μ are damping functions given by [5]:

$$f_2 = \left\{1 - exp\left[-\frac{(v\varepsilon)^{0.25}n}{3.1\,v}\right]\right\}^2 \times \left\{1 - 0.3exp\left[-\left(\frac{(k^2/v\varepsilon)}{6.5}\right)^2\right]\right\} \quad (2.18)$$

$$f_\mu = \left\{1 - exp\left[-\frac{(v\varepsilon)^{0.25}n}{14\,v}\right]\right\}^2 \times \left\{1 + \frac{5}{(k^2/v\varepsilon)^{0.75}}exp\left[-\left(\frac{(k^2/v\varepsilon)}{200}\right)^2\right]\right\} \quad (2.19)$$

where n is the coordinate normal to the wall.

2.4 One-Energy Equation Model

Similarly, macroscopic energy equations are obtained for both fluid and solid phases by applying time and volume average operators to Eqs. (2.3) and (2.4). As in the flow case, the integrations are performed over a Representative Elementary Volume (REV), resulting in,

$$(\rho c_p)_f \left[\frac{\partial \phi \langle \overline{T_f} \rangle^i}{\partial t} + \nabla \cdot \left\{\phi \left(\langle \bar{\mathbf{u}} \rangle^i \langle \overline{T_f} \rangle^i + \underbrace{\langle {}^i\bar{\mathbf{u}} {}^i\overline{T_f} \rangle^i}_{\text{thermaldisperson}} + \underbrace{\overline{\langle \mathbf{u}' \rangle^i \langle T_f' \rangle^i}}_{\substack{\text{turbulent heat} \\ \text{flux}}} + \underbrace{\overline{\langle {}^i\mathbf{u}' {}^iT_f' \rangle^i}}_{\substack{\text{turbulentthermal} \\ \text{disperson}}}\right)\right\}\right] \quad (2.20)$$

where A_i is the interfacial area between phases and the expansion,

$$\langle \overline{\mathbf{u}'T_f'} \rangle^i = \langle \overline{(\langle \mathbf{u}' \rangle^i + {}^i\mathbf{u}')\,(\langle T_f' \rangle^i + {}^iT')} \rangle^i = \overline{\langle \mathbf{u}' \rangle^i \langle T_f' \rangle^i} + \langle \overline{{}^i\mathbf{u}'\,{}^iT_f'} \rangle^i \quad (2.21)$$

has been used in light of the double decomposition concept given by Eqs. (2.7)–(2.10) [2]. For the solid phase, one has,

$$(\rho c_p)_s \left\{\frac{\partial (1-\phi)\langle \overline{T_s} \rangle^i}{\partial t}\right\} = \nabla \cdot \underbrace{\left\{k_s \nabla[(1-\phi)\langle \overline{T_s} \rangle^i] - \frac{1}{\Delta V}\int_{A_i} \mathbf{n}_i k_s \overline{T_s}\,dA\right\}}_{\text{conduction}}$$

$$\underbrace{-\frac{1}{\Delta V}\int_{A_i} \mathbf{n}_i \cdot k_s \nabla \overline{T_s}\,dA}_{\text{interfacial heat transfer}} \quad (2.22)$$

An interfacial heat transfer coefficient is needed when corresponding terms in Eqs. (2.20) and (2.22) are modeled following the local thermal non-equilibrium assumption [6]. Here, however, we assume local thermal equilibrium between the fluid and solid phases, i.e., we add Eqs. (2.20) and (2.22) and consider $\langle \overline{T}_f \rangle^i = \langle \overline{T}_s \rangle^i = \langle \overline{T} \rangle^i$, giving further:

$$\left\{ (\rho c_p)_f \phi + (\rho c_p)_s (1 - \phi) \right\} \frac{\partial \langle \overline{T} \rangle^i}{\partial t} + (\rho c_p)_f \nabla \cdot (\mathbf{u}_D \langle \overline{T} \rangle^i)$$

$$= \nabla \cdot \left\{ [k_f \phi + k_s (1 - \phi)] \nabla \langle \overline{T} \rangle^i \right\} + \nabla \cdot \left[\frac{1}{\Delta V} \int_{A_i} \mathbf{n} (k_f \overline{T}_f - k_s \overline{T}_s) dS \right]$$

$$- (\rho c_p)_f \nabla \cdot \left[\phi \left(\langle {}^i \overline{\mathbf{u}}^i \overline{T}_f \rangle^i + \langle \overline{\mathbf{u}' T'_f} \rangle^i \right) \right] \tag{2.23}$$

The interface conditions at A_i are further given by,

$$\left. \begin{array}{c} T_f = T_s \\ \mathbf{n} \cdot (k_f \nabla T_f) = \mathbf{n} \cdot (k_s \nabla T_s) \end{array} \right\} \quad in \ A_i \tag{2.24}$$

Equation (2.23) express the one-equation model for heat transport in porous media. Further, using the double decomposition concept, Rocamora and de Lemos [2] have shown that the last term on the right hand side of Eq. (2.23) can be expressed as:

$$\langle \overline{\mathbf{u}' T'_f} \rangle^i = \langle \overline{(\langle \mathbf{u}' \rangle^i + {}^i \mathbf{u}')(\langle T'_f \rangle^i + {}^i T')} \rangle^i = \overline{\langle \mathbf{u}' \rangle^i \langle T'_f \rangle^i} + \langle \overline{{}^i \mathbf{u}'{}^i T'_f} \rangle^i \tag{2.25}$$

So, in view of Eqs. (2.21) and (2.23) can be rewritten as:

$$\left\{ (\rho c_p)_f \phi + (\rho c_p)_s (1 - \phi) \right\} \frac{\partial \langle \overline{T} \rangle^i}{\partial t} + (\rho c_p)_f \nabla \cdot (\mathbf{u}_D \langle \overline{T} \rangle^i)$$

$$= \nabla \cdot \left\{ [k_f \phi + k_s (1 - \phi)] \nabla \langle \overline{T} \rangle^i \right\} + \nabla \cdot \underbrace{\left[\frac{1}{\Delta V} \int_{A_i} \mathbf{n} (k_f \overline{T}_f - k_s \overline{T}_s) dS \right]}_{I}$$

$$- (\rho c_p)_f \nabla \cdot \left[\phi \left(\underbrace{\overline{\langle \mathbf{u}' \rangle^i \langle T'_f \rangle^i}}_{II} + \underbrace{\langle {}^i \overline{\mathbf{u}}^i \overline{T}_f \rangle^i}_{III} + \underbrace{\langle \overline{{}^i \mathbf{u}'{}^i T'_f} \rangle^i}_{IV} \right) \right] \tag{2.6}$$

where to the underscored terms in Eq. (2.26) the following physical significance can be attributed:

I. Tortuosity based on microscopic time averaged temperatures.
II. Turbulent heat flux due to the fluctuating components of macroscopic velocity and temperature $(\overline{\langle \mathbf{u}' \rangle^i \langle T'_f \rangle^i} = \overline{\langle \mathbf{u} \rangle^{i'} \langle T_f \rangle^{i'}})$.

III. Thermal dispersion associated with deviations of microscopic time average velocity and temperature. Note that this term is also present when analyzing laminar convective heat transfer in porous media.
IV. Turbulent thermal dispersion in a porous medium due to both time fluctuations and spatial deviations of both microscopic velocity and temperature.

In order to apply Eq. (2.26) to obtain the temperature field for turbulent flow in porous media, the underscored terms have to be modeled in some way as a function of the surface average temperature, $\langle \bar{T} \rangle^i$. To accomplish this, a gradient type diffusion model is used for all the terms, i.e., tortuosity (I), turbulent heat flux due to temporal fluctuations (II), thermal dispersion due to spatial deviations (III) and turbulent thermal dispersion due to temporal fluctuations and spatial deviations (IV).

Using these gradient type diffusion models, we can write:

$$\text{Tortuosity:}\ \left[\frac{1}{\Delta V} \int_{A_i} \mathbf{n} \left(k_f \overline{T_f} - k_s \overline{T_s} \right) dS \right] = \mathbf{K}_{tor} \cdot \nabla \langle \bar{T} \rangle^i \qquad (2.27)$$

$$\text{Turbulent heat flux:}\ - \left(\rho c_p \right)_f \left(\overline{\phi \langle \mathbf{u}' \rangle^i \langle T_f' \rangle^i} \right) = \mathbf{K}_t \cdot \nabla \langle \bar{T} \rangle^i \qquad (2.28)$$

$$\text{Thermal dispersion:}\ - \left(\rho c_p \right)_f \left(\phi \langle {}^i \bar{\mathbf{u}}^i \overline{T_f} \rangle^i \right) = \mathbf{K}_{disp} \cdot \nabla \langle \bar{T} \rangle^i \qquad (2.29)$$

$$\text{Turbulent thermal dispersion:}\ - \left(\rho c_p \right)_f \left(\phi \langle {}^i \overline{\mathbf{u}'^i T_f'} \rangle^i \right) = \mathbf{K}_{disp,t} \cdot \nabla \langle \bar{T} \rangle^i \qquad (2.30)$$

For the above shown expressions, Eq. (2.26) can be rewritten as:

$$\left\{ \left(\rho c_p \right)_f \phi + \left(\rho c_p \right)_s (1 - \phi) \right\} \frac{\partial \langle \bar{T} \rangle^i}{\partial t} + \left(\rho c_p \right)_f \nabla \cdot \left(\mathbf{u}_D \langle \bar{T} \rangle^i \right) = \nabla \cdot \left\{ \mathbf{K}_{eff} \cdot \nabla \langle \bar{T} \rangle^i \right\}$$
$$(2.31)$$

where, $\mathbf{K}_{eff}$, given by:

$$\mathbf{K}_{eff} = \underbrace{\left[\phi k_f + (1 - \phi) k_s \right]}_{k_{eff}} \mathbf{I} + \mathbf{K}_{tor} + \mathbf{K}_t + \mathbf{K}_{disp} + \mathbf{K}_{disp,t} \qquad (2.32)$$

is the effective overall conductivity tensor.

In order to be able to apply Eq. (2.31), it is necessary to determine the conductivity tensors in Eq. (2.32), i.e., $\mathbf{K}_{tor}$, $\mathbf{K}_t$, $\mathbf{K}_{disp}$ and $\mathbf{K}_{disp,t}$. Following Kuwahara and Nakayama [7], this can be accomplished for the tortuosity and thermal dispersion conductivity tensors, $\mathbf{K}_{tor}$ and $\mathbf{K}_{disp}$, by making use of a unit cell subjected to periodic boundary conditions for the flow and a linear temperature gradient imposed over the domain. The conductivity tensors are then obtained directly from the microscopic results (see [8] for detail). Nevertheless, for simplicity, the tortuosity and dispersion mechanisms are here neglected.

The turbulent heat flux and turbulent thermal dispersion terms, $\mathbf{K}_t$ and $\mathbf{K}_{disp,t}$, which can not be determined from such a microscopic calculation, are modeled through the Eddy diffusivity concept, similarly to Kuwahara and Nakayama [7]. It should be noticed that these terms arise only if the flow is turbulent, whereas the tortuosity and the thermal dispersion terms exist for both laminar and turbulent flow regimes.

The macroscopic version of the 'turbulent heat flux' is given by:

$$-\left(\rho c_p\right)_f \langle \mathbf{u}'T_f' \rangle^i = \left(\rho c_p\right)_f \frac{v_{t_\phi}}{\sigma_{t_\phi}} \nabla \langle \bar{T}_f \rangle^i \tag{2.33}$$

where v_{t_ϕ} is the macroscopic cinematic eddy viscosity related to dynamic Eddy viscosity, $\mu_{t_\phi} = \rho v_{t_\phi}$ given by Eq. (2.15) and $\sigma_{t_\phi} = 0.9$ is the macroscopic turbulent Prandtl number.

According to Eqs. (2.21) and (2.33), the macroscopic heat flux due to turbulence is taken as the sum of the turbulent heat flux and the turbulent thermal dispersion found by Rocamora and de Lemos [2]. In view of the arguments given above, the turbulent heat flux and turbulent thermal dispersion components of the conductivity tensor, $\mathbf{K}_t$ and $\mathbf{K}_{disp,t}$, respectively, will be expressed as:

$$\mathbf{K}_t + \mathbf{K}_{disp,t} = \phi (\rho c_p)_f \frac{v_{t_\phi}}{\sigma_{t_\phi}} \mathbf{I} \tag{2.34}$$

2.5 Two-Energy Equation Model

Similarly, macroscopic energy equations are obtained for both fluid and solid phases by applying time and volume average operators to local instantaneous energy equations, resulting in,

$$(\rho c_p)_f \left[\frac{\partial \phi \langle \bar{T}_f \rangle^i}{\partial t} + \nabla \cdot \left\{ \phi \left(\langle \bar{\mathbf{u}} \rangle^i \langle \bar{T}_f \rangle^i + \underbrace{\langle {}^i \bar{\mathbf{u}}{}^i \overline{T_f} \rangle^i}_{\text{thermal disperson}} + \underbrace{\langle \mathbf{u}' \rangle^i \langle T_f' \rangle^i}_{\substack{\text{turbulent heat} \\ \text{flux}}} + \underbrace{\langle {}^i \mathbf{u}'{}^i T_f' \rangle^i}_{\substack{\text{turbulent thermal} \\ \text{disperson}}} \right) \right\} \right]$$

$$= \nabla \cdot \underbrace{\left[k_f \nabla \left(\phi \langle \bar{T}_f \rangle^i \right) + \frac{1}{\Delta V} \int_{A_i} \mathbf{n}_i k_f \overline{T}_f dA \right]}_{\text{conduction}} + \underbrace{\frac{1}{\Delta V} \int_{A_i} \mathbf{n}_i \cdot k_f \nabla \overline{T}_f dA}_{\text{interfacial heat transfer}}$$

where the expansion,

$$\langle \overline{\mathbf{u}'T_f'} \rangle^i = \overline{\langle (\langle \mathbf{u}' \rangle^i + {}^i \mathbf{u}')(\langle T_f' \rangle^i + {}^i T') \rangle^i} = \overline{\langle \mathbf{u}' \rangle^i \langle T_f' \rangle^i} + \langle \overline{{}^i \mathbf{u}'{}^i T_f'} \rangle^i \tag{2.36}$$

has been used in light of the double decomposition concept given by [2]. For the solid phase, one has,

$$
(\rho c_p)_s \left\{ \frac{\partial (1-\phi)\langle \overline{T_s} \rangle^i}{\partial t} \right\} = \nabla \cdot \underbrace{\left\{ k_s \nabla \left[(1-\phi)\langle \overline{T_s} \rangle^i \right] - \frac{1}{\Delta V} \int_{A_i} \mathbf{n}_i k_s \overline{T_s} dA \right\}}_{\text{conduction}}
$$

$$
\underbrace{- \frac{1}{\Delta V} \int_{A_i} \mathbf{n}_i \cdot k_s \nabla \overline{T_s} dA}_{\text{interfacial heat transfer}}
\tag{2.37}
$$

In (2.20) and (2.22), $\langle \overline{T_s} \rangle^i$ and $\langle \overline{T_f} \rangle^i$ denote the intrinsic average temperature of solid and fluid phases, respectively, A_i is the interfacial area within the REV and $\mathbf{n}_i$ is the unit vector normal to the fluid–solid interface, pointing from the fluid towards the solid phase. Equations (2.20) and (2.22) are the macroscopic energy equations for the fluid and the porous matrix (solid), respectively.

In order to use Eqs. (2.20) and (2.22), the underscored terms have to be modeled in some way as a function of the intrinsically averaged temperature of solid phase and fluid, $\langle \overline{T_s} \rangle^i$ and $\langle \overline{T_f} \rangle^i$. To accomplish this, a gradient type diffusion model is used for all the terms, in the form,

$$
\text{Turbulent heat flux: } - (\rho c_p)_f \left(\phi \overline{\langle \mathbf{u}' \rangle^i \langle T_f' \rangle^i} \right) = \mathbf{K}_t \cdot \nabla \langle \overline{T}_f \rangle^i
\tag{2.38}
$$

$$
\text{Thermal dispersion: } - (\rho c_p)_f \left(\phi \langle {}^i \overline{\mathbf{u}}\, \overline{T_f} \rangle^i \right) = \mathbf{K}_{disp} \cdot \nabla \langle \overline{T}_f \rangle^i
\tag{2.39}
$$

$$
\text{Turbulent thermal dispersion: } - (\rho c_p)_f \left(\phi \langle {}^i \mathbf{u}'^i T_f' \rangle^i \right) = \mathbf{K}_{disp,t} \cdot \nabla \langle \overline{T}_f \rangle^i
\tag{2.40}
$$

$$
\text{Local conduction: } \nabla \cdot \left[\frac{1}{\Delta V} \int_{A_i} \mathbf{n}_i k_f \overline{T_f} dA \right] = \mathbf{K}_{f,s} \cdot \nabla \langle \overline{T}_s \rangle^i
$$

$$
- \nabla \cdot \left[\frac{1}{\Delta V} \int_{A_i} \mathbf{n}_i k_s \overline{T_s} dA \right] = \mathbf{K}_{s,f} \cdot \nabla \langle \overline{T}_f \rangle^i
\tag{2.41}
$$

where $\mathbf{n}_i$ in (2.27) is the unit vector pointing outwards of the fluid phase. In this work, for simplicity, one assumed that for turbulent flow the overall thermal resistance between the two phases is controlled by the interfacial film coefficient, rather than by the thermal resistance within each phase. As such, the tortuosity coefficients $\mathbf{K}_{f,s}, \mathbf{K}_{s,f}$ are here neglected for the sake of simplicity.

The heat transferred between the two phases can be modeled by means of a film coefficient h_i such that,

$$h_i a_i \left(\langle \overline{T_s} \rangle^i - \langle \overline{T_f} \rangle^i \right) = \frac{1}{\Delta V} \int_{A_i} \mathbf{n}_i \cdot k_f \nabla \overline{T_f} \, dA = \frac{1}{\Delta V} \int_{A_i} \mathbf{n}_i \cdot k_s \nabla \overline{T_s} \, dA \qquad (2.42)$$

where $a_i = A_i / \Delta V$ is the surface area per unit volume.

For the above shown expressions, Eqs. (2.20) and (2.22) can be written as:

$$\left\{ (\rho c_p)_f \phi \right\} \frac{\partial \langle \overline{T_f} \rangle^i}{\partial t} + (\rho c_p)_f \nabla \cdot \left(\mathbf{u}_D \langle \overline{T_f} \rangle^i \right) = \nabla \cdot \left\{ \mathbf{K}_{eff,f} \cdot \nabla \langle \overline{T_f} \rangle^i \right\} \\ + h_i a_i \left(\langle \overline{T_s} \rangle^i - \langle \overline{T_f} \rangle^i \right) \qquad (2.43)$$

$$\left\{ (1 - \phi)(\rho c_p)_s \right\} \frac{\partial \langle \overline{T_s} \rangle^i}{\partial t} = \nabla \cdot \left\{ \mathbf{K}_{eff,s} \cdot \nabla \langle \overline{T_s} \rangle^i \right\} - h_i a_i \left(\langle \overline{T_s} \rangle^i - \langle \overline{T_f} \rangle^i \right) \qquad (2.44)$$

where, $\mathbf{K}_{eff,f}$ and $\mathbf{K}_{eff,s}$ are the effective conductivity tensor for fluid and solid, respectively, given by:

$$\mathbf{K}_{eff,f} = [\phi k_f]\mathbf{I} + \mathbf{K}_{f,s} + \mathbf{K}_t + \mathbf{K}_{disp} + \mathbf{K}_{disp,t} \qquad (2.45)$$

$$\mathbf{K}_{eff,s} = [(1 - \phi)k_s]\mathbf{I} + \mathbf{K}_{s,f} \qquad (2.46)$$

and $\mathbf{I}$ is the unit tensor.

2.5.1 Turbulence

In order to apply Eqs. (2.31)–(2.44) for the Two-Energy Equation Model (2EEM), or Eq. (2.31) for the One-Energy Equation Model (1EEM), it is necessary to determine the components of the effective conductivity tensors in Eq. (2.32), i.e., $\mathbf{K}_{f,s}$, $\mathbf{K}_t$, $\mathbf{K}_{disp}$ and $\mathbf{K}_{disp,t}$ for the 2EEM, and $\mathbf{K}_{tor}$, $\mathbf{K}_t$, $\mathbf{K}_{disp}$ and $\mathbf{K}_{disp,t}$ in Eq. (2.32), for the 1EEM. Here, for simplicity, all mechanisms are neglected except turbulence, which is here explicitly accounted for (see [7, 8] for a discussion on the determination of such coefficients).

The turbulent heat flux and turbulent thermal dispersion terms, $\mathbf{K}_t$ and $\mathbf{K}_{disp,t}$, are here modeled through the Eddy diffusivity concept, as:

$$\mathbf{K}_t + \mathbf{K}_{disp,t} = \phi (\rho c_p)_f \frac{\nu_{t_\phi}}{\sigma_T} \mathbf{I} \qquad (2.47)$$

2.5.2 Interfacial Heat Transfer, h_i

Wakao et al. [9] proposed a correlation for h_i for closely packed bed and compared results with their experimental data. This correlation reads,

$$\frac{h_i D}{k_f} = 2 + 1.1 Re_D^{0.6} Pr^{1/3}. \tag{2.48}$$

Kuwahara et al. [10] also obtained the interfacial convective heat transfer coefficient for laminar flow, as follows,

$$\frac{h_i D}{k_f} = \left(1 + \frac{4(1-\phi)}{\phi}\right) + \frac{1}{2}(1-\phi)^{1/2} Re_D \overset{1/3}{Pr}, \text{ valid for } 0.2 < \phi < 0.9 \tag{2.49}$$

Equation (2.49) is based on porosity dependency and is valid for packed beds of particle diameter D. Following this same methodology, in which the porous medium is considered an infinite number of solid square rods, Saito and de Lemos [6] proposed a correlation for obtaining the interfacial heat transfer coefficient for turbulent flow as,

$$\frac{h_i D}{k_f} = 0.08 \left(\frac{Re_D}{\phi}\right)^{0.8} Pr^{1/3}; \text{ for } 1.0 \times 10^4 < \frac{Re_D}{\phi} < 2.0 \times 10^7, \tag{2.50}$$
$$\text{valid for } 0.2 < \phi < 0.9,$$

2.5.3 Non-Dimensional Parameters

The local Nusselt number for the one-energy equation model used by [11] is given by:

$$Nu = \left(\frac{\partial \langle \bar{T} \rangle^i}{\partial y}\right)_{y=H} \frac{H}{T_1 - T_0} \tag{2.51}$$

Equation (2.51) assumes the local thermal equilibrium hypothesis, i.e., $\langle T \rangle^i = \langle T_s \rangle^i = \langle T_f \rangle^i$. When the Local Non-thermal Equilibrium model is applied, that are distinct definitions for the Nusselt number associated to each phase, as follows [12],

Fluid phase Nusselt number:

$$Nu_f = \left(\frac{\partial \langle \bar{T}_f \rangle^i}{\partial y}\right)_{y=H} \frac{H}{T_1 - T_0} \tag{2.52}$$

Solid phase Nusselt number:

$$Nu_s = \left(\frac{\partial \langle \bar{T}_s \rangle^i}{\partial y}\right)_{y=H} \frac{H}{T_1 - T_0} \tag{2.53}$$

2.6 Boundary Conditions and Numerical Details

2.6.1 Flow Boundary and Interface Conditions

At the interface between the porous layer and the clear region the macroscopic velocity, intrinsic pressure, turbulence kinetic energy and its dissipation rate, as well as their respective diffusive fluxes, were assumed to be continuous functions so that,

$$\bar{\mathbf{u}}_D|_{0<\phi<1} = \bar{\mathbf{u}}_D|_{\phi=1} \tag{2.54}$$

$$\langle \bar{p} \rangle^i|_{0<\phi<1} = \langle \bar{p} \rangle^i|_{\phi=1} \tag{2.55}$$

$$\langle k \rangle^v|_{0<\phi<1} = \langle k \rangle^v|_{\phi=1} \tag{2.56}$$

$$\left(\mu + \frac{\mu_{t_\phi}}{\sigma_k}\right) \frac{\partial \langle k \rangle^v}{\partial y}\bigg|_{0<\phi<1} = \left(\mu + \frac{\mu_t}{\sigma_k}\right) \frac{\partial \langle k \rangle^v}{\partial y}\bigg|_{\phi=1} \tag{2.57}$$

$$\langle \varepsilon \rangle^v|_{0<\phi<1} = \langle \varepsilon \rangle^v|_{\phi=1} \tag{2.58}$$

$$\left(\mu + \frac{\mu_{t_\phi}}{\sigma_\varepsilon}\right) \frac{\partial \langle \varepsilon \rangle^v}{\partial y}\bigg|_{0<\phi<1} = \left(\mu + \frac{\mu_t}{\sigma_\varepsilon}\right) \frac{\partial \langle \varepsilon \rangle^v}{\partial y}\bigg|_{\phi=1} \tag{2.59}$$

2.6.2 Heat Boundary and Interface Conditions

For the planar geometries in Fig. 1.2, temperature of the incoming fluid was specified at jet inlet. At the flow exit ($x = L$), a zero diffusion flux condition was imposed and a symmetry plane was considered at the left boundary ($x = 0$). On the bottom wall, a constant given temperature T_1 was assumed and, on the top wall, thermal isolation condition prevailed.

At the macroscopic interface shown in Fig. 1.2b ($y = h$), the volume-time-averaged fluid temperature was continuous as well as the total transverse heat flux (laminar plus turbulent). As a consequence of the imposed continuity of the fluid phase heat flux at the interface, the heat conducted by the solid matrix in the *y*-direction at the macroscopic boundary was of null value. As will be shown in the chapters to follow, solid temperature gradients attained zero value at $y = h$.

2.6.3 Numerical Details

Equations (2.11), (2.12) and (2.31), subject to interface and boundary conditions were discretized in a two-dimensional control volume involving both clear and porous media. The finite volume method was used for discretizing the equation set and the SIMPLE algorithm was applied to handle the pressure-velocity coupling (Patankar [13]). The discretized form of the two-dimensional conservation equation for a generic property φ in permanent regime reads,

$$I_e + I_w + I_n + I_s = S_\varphi \qquad (2.60)$$

where I_e, I_w, I_n and I_s represent, respectively, the fluxes of φ in the faces east, west, north and south faces of the control volume and S_φ its term source.

Standard source term linearization is accomplished by using,

$$S_\varphi \approx S_\varphi^{**} \langle \varphi \rangle_P^i + S_\varphi^* \qquad (2.61)$$

Discretization in the *x*-direction momentum equation gives,

$$S^{*x} = \left(S_e^{*x} \right)_P - \left(S_w^{*x} \right)_P + \left(S_n^{*x} \right)_P - \left(S_s^{*x} \right)_P + S_P^* \qquad (2.62)$$

$$S^{**x} = S_\phi^{**} \qquad (2.63)$$

where, S^{*x} is the diffusive part, here treated in an explicit form. The second term, S^{**x}, entails the additional drag forces due to the porous matrix, last two terms in Eq. (2.12), are here treated explicitly.

References

1. D.R. Graminho, M.J.S. de Lemos, Simulation of turbulent impinging jet into a cylindrical chamber with and without a porous layer at the bottom. Int. J. Heat Mass Transf. **52**, 680–693 (2009)
2. F.D. Rocamora Jr, M.J.S. de Lemos, Analysis of convective heat transfer of turbulent flow in saturated porous media. Int. Commun. Heat Mass Transf. **27**(6), 825–834 (2000)
3. W.G. Gray, P.C.Y. Lee, On the theorems for local volume averaging of multiphase system. Int. J. Multiph. Flow **3**(4), 333–340 (1977)
4. M.J.S. de Lemos, *Turbulence in Porous Media: Modeling and Applications* (Elsevier, Amsterdam, 2006)
5. K. Abe, Y. Nagano, T. Kondoh, An improved k-ε model for prediction of turbulent flows with separation and reattachment. Trans. JSME Ser. B **58**(554), 3003–3010 (1992)

6. M.B. Saito, M.J.S. de Lemos, A correlation for interfacial heat transfer coefficient for turbulent flow over an array of square rods. J. Heat Transf. **128**, 444–452 (2006)
7. F. Kuwahara, A. Nakayama, Numerical modeling of non-Darcy convective flow in a porous medium, in *Proceedings 11th International Heat Transfer Conference*, vol. 4 (Kyongyu, Korea, 1998), pp. 411–416
8. A. Nakayama, F. Kuwahara, A macroscopic turbulence model for flow in a porous medium. J. Fluids Eng. **121**, 427–433 (1999)
9. N. Wakao, S. Kaguei, T. Funazkri, Effect of fluid dispersion coefficients on particle-to-fluid heat transfer coefficients in packed bed. Chem. Eng. Sci. **34**, 325–336 (1979)
10. F. Kuwahara, M. Shirota, A. Nakayama, A numerical study of interfacial convective heat transfer coefficient in two-energy equation model for convection in porous media. Int. J. Heat Mass Transf. **44**, 1153–1159 (2001)
11. M.J.S. de Lemos, C. Fischer, Thermal analysis of an impinging jet on a plate with and without a porous layer. Numer. Heat Transf. Part A Appl. **54**, 1022–1041 (2008)
12. B. Alazmi, K. Vafai, Analysis of variants within the porous media transport models. J. Heat Transf. **122**, 303–326 (2000)
13. S.V. Patankar, *Numerical Heat Transfer and Fluid Flow* (Hemisphere, New York, 1980)

Chapter 3
Flow Structure of Impinging Jets

First, for code validation, initial simulations were conducted in the clear chamber of height H, where free flow occurs (Fig. 1.1b). This first set of simulations considers that a solid wall is located at depth H and no porous layer is positioned at the bottom of the chamber. Therefore, streamlines, velocity profiles and turbulence kinetic energy contours for an empty enclosure are presented prior to showing computations considering the porous layer in Fig. 1.1a. Also, for convenience in presenting results, all figures below, which contain streamline patterns, are plotted with the symmetry axis aligned with the horizontal direction, with the jet coming at the lower left corner of the figures.

Also, it is important to emphasize the main differences of the work herein and that presented by Prakash et al. [1]. Here, the turbulence model used is composed by only one set of equations and is applied without any distinction within the fluid layer and the porous material. These equations are (2.11), (2.12), (2.16), (2.17). In Prakash et al. [1], the model of Antohe and Lage [2] was applied, which is based on a volume-time sequence of integration of local instantaneous equations. In addition, more than one set of equations were used in [1], with different combinations of distinct terms in their modeled k-equation. Besides, a closure where the flow was assumed laminar within the porous matrix was also considered in [1]. Here, only one model is employed to handle turbulence, regardless if the computation node lies in the free flow region or in the porous matrix.

Further, the mathematical closure here employed follows the work of Pedras and de Lemos [3], which used the reversed order of integration (time-volume sequence) for obtaining the set of transport equations. The relationship between these two models is explained by a new concept named double-decomposition. The interested reader is referred to de Lemos [4] where more information on such concept can be found.

Therefore, besides the use of a different numerical technique, with one single computation grid for the entire computational domain, the work herein makes use of a model that needs no adjustments if it is applied in free flow regions (fluid

M. J. S. de Lemos, *Turbulent Impinging Jets into Porous Materials*,   21
SpringerBriefs in Computational Mechanics,
DOI: 10.1007/978-3-642-28276-8_3, © The Author(s) 2012

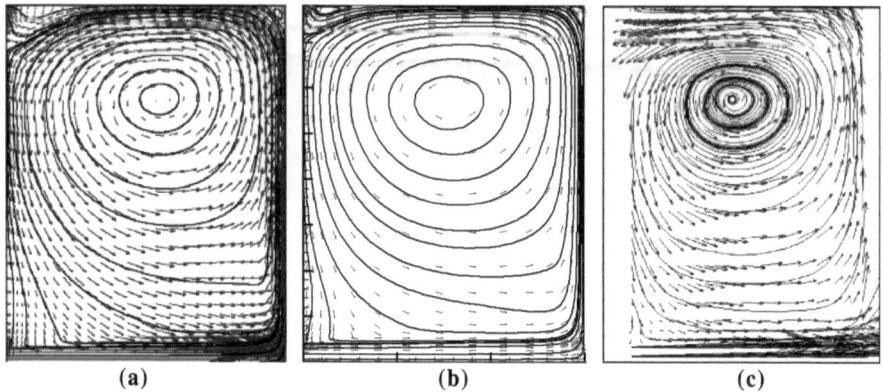

Fig. 3.1 Comparison of streamfunctions for $Re = 30,000$, $H = 0.15$ m: **a** CFD results—Prakash et al. [1, 5], **b** present results, **c** LDV measurements—Prakash et al. [1, 5]

layer) or in the porous substrate. Having the distinction between such two approaches been clarified, the sections below present simulations for a clear chamber and for the case where a porous material is positioned at the bottom of the cylinder shown in Fig. 1.1.

3.1 Clear Chamber

3.1.1 Mean Flow

Figure 3.1 shows a comparison of the numerically simulated streamfunctions with experimental flow visualizations and CFD results from Prakash et al. [1] for $H = 0.15$ m. For the present case, the entire half plane of the cylinder is filled with one large recirculation cell. A small secondary recirculation bubble can be seem in the top-right corner of the picture, but with negligible dimensions compared to the main one. Simulations seem to be close to literature results, with main flow streamlines following very closely the pattern given by the LDV experiments. Also, the flow in the core region and close to the walls agrees well with available CFD results, indicating that simulations herein may well represent actual flow behavior within the entire chamber.

The effect of the fluid layer height, H, given in Fig. 3.2 shows the stream-functions for $H = 0.15$, 0.10 and 0.05 m. By decreasing the liquid height, main flow behavior remains the same, with a central dominant recirculation. The peripheral recirculations get smaller with the decrease in H. For $H = 0.1$ m, a recirculation appears close to the jet exit, and both secondary recirculations close to the cylinder wall are reduced compared to the case with $H = 0.15$ m. With a

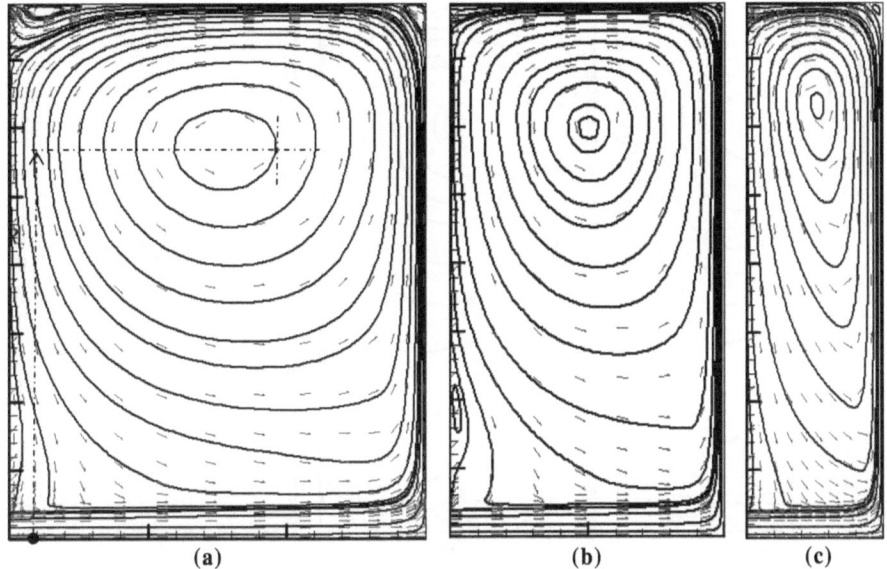

(a)   (b)   (c)

Fig. 3.2 Effect of fluid layer height on streamlines for $Re = 30,000$. **a** $H = 0.15$ m, **b** $H = 0.10$ m, **c** $H = 0.05$ m

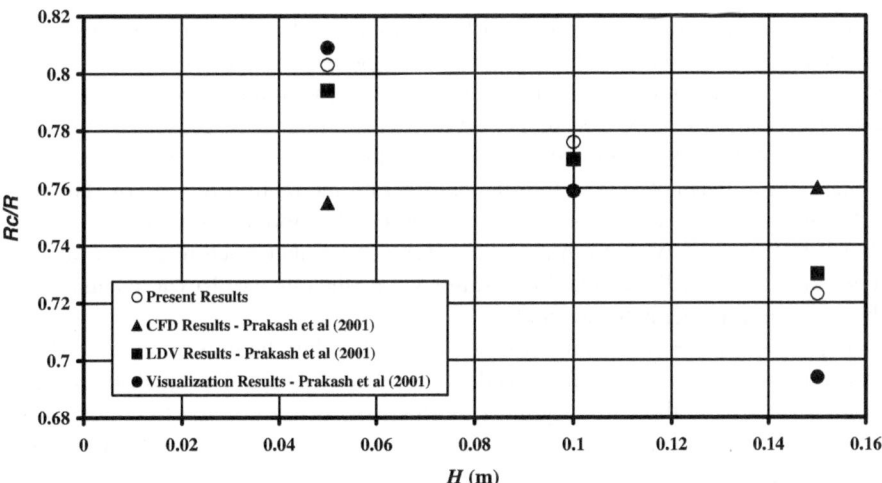

Fig. 3.3 Radial position of center recirculation as a function of fluid layer height, H

further decrease in H, the main recirculation becomes elongated and the peripheral ones nearly vanish.

Figure 3.3 shows the horizontal position of the main recirculation for different liquid heights. The R_c/R axis represents the normalized radial position of the

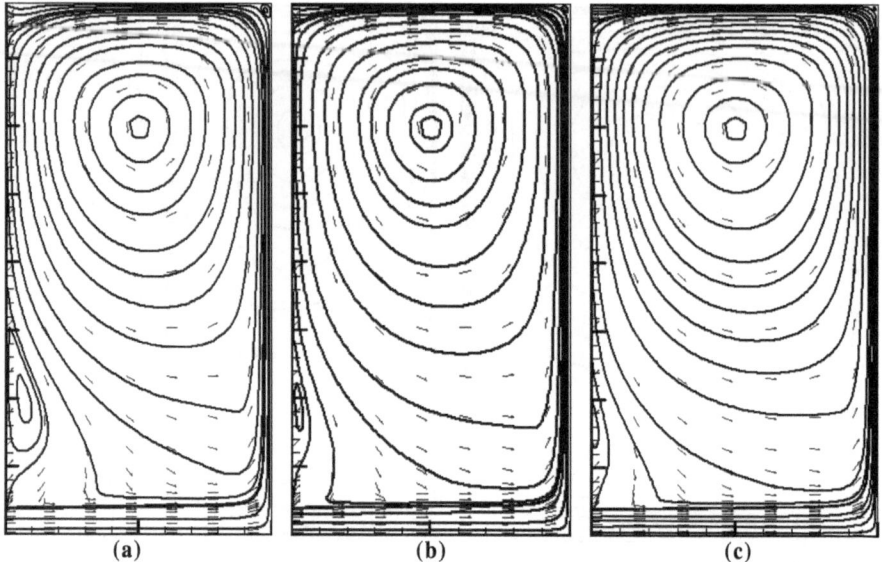

Fig. 3.4 Effect of Reynolds number on streamfunctions for $H = 0.10$ m: **a** $Re = 18{,}900$, **b** $Re = 30{,}000$, **c** $Re = 47{,}000$

recirculation, where R is the radius of the cylindrical chamber. From the picture, it can be seen that the center of the recirculation moves closer to the cylinder wall as H decreases. Present computations seem to follow the trends in both measurement and visualization results, i.e., reduction of R_c as H increases. Figure 3.4 gives the influence of Reynolds number on the main flow pattern. It can be seen that a change in Re does not have a significant effect on the main flow pattern. The most noticeable change is a secondary recirculation developing close to the jet exit (lower left corner). With an increase in Reynolds number, such recirculation tends to get smaller up to $Re = 30{,}000$. Past this value, the effect of Reynolds number on the flow pattern seems to be small.

Figure 3.5 shows the axial velocity profiles for fluid heights equal to 0.15, 0.1 and 0.05 m. In this figure, the profiles have been plotted for six different axial locations covering the entire cylindrical chamber. It can be seen that close to the jet centerline and near the surface at the bottom, the present results seem to fall in between numerical and experimental data available in the literature. At higher axial distance, present simulations align with CFD predictions in the literature, but both overpredict the axial velocity at the centerline. According to Prakash et al. [5], experimental conditions at inlet did not correspond to fully developed flow, whereas numerical simulations here reported assumed such condition. In general, the present simulations fail to predict the reduction in the centerline velocity as the bottom surface approaches. However, elsewhere in the chamber the main features of the flow were reproduced.

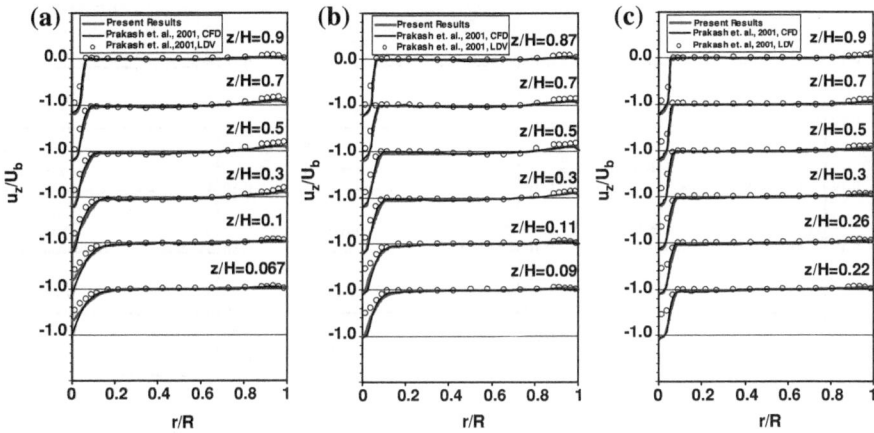

Fig. 3.5 Axial velocity profiles for cases without foam: **a** $H = 0.15$ m, **b** $H = 0.10$ m, **c** $H = 0.05$ m

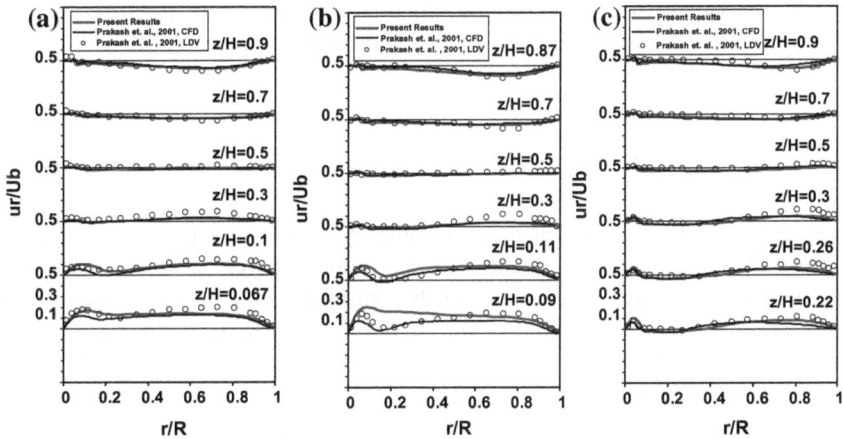

Fig. 3.6 Radial velocity profiles for cases without foam: **a** $H = 0.15$ m, **b** $H = 0.10$ m, **c** $H = 0.05$ m

Radial velocity profiles for the three fluid layers investigated are presented in Fig. 3.6. For $H = 0.15$, results follow reasonably well experimental data for $z/H > 0.5$, but underpredicts measured velocities as one gets closer to the plate at the bottom. The same trend is observed in the reported CFD predictions. For $H = 0.10$ and $r/R < 0.1$, the present computations reproduce the experimental data reasonably well, but in the range $0.1 < r/R < 0.6$ and for lower values of z/H, simulations herein fall above experimental values for velocities. At $H = 0.05$, the present curves seem to closely follow the experimental data trend for higher values of z/H. As the flow moves away from the symmetry axis and further down from the

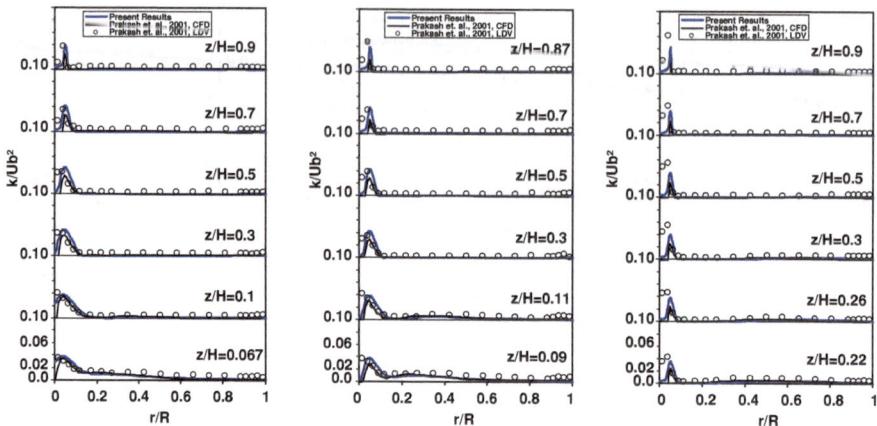

Fig. 3.7 Turbulence kinetic energy profiles for cases without foam: **a** $H = 0.15$ m, **b** $H = 0.10$ m, **c** $H = 0.05$ m

jet exit, the quality of predictions seem to deteriorate. Nevertheless, the main features of the flow were captured by the simulations reported here.

3.1.2 Turbulent Field

Figure 3.7 shows the turbulence kinetic energy profiles for the same conditions as above. For all fluid layer heights, results reproduce very closely experimental data, including the peak value of k near the symmetry axis, which is caused by the strong shear layer formed around the incoming jet. Close to the impingement plate, a second elevation in turbulence kinetic energy appears. This second maximum is attributed to a laminar-turbulent boundary layer transition in the wall jet region, as defined in Incropera and DeWitt [6]. That is, the flow decelerates when coming close to the wall, turbulence is damped and the laminar regime prevails in the stagnation region. Afterwards, the flow turns and accelerates along the wall. Overall, simulations herein seem to reproduce basic aspects of the flow pattern, providing the necessary code validation prior to presenting next set of simulations, which include a layer of porous material at the bottom of the chamber.

Before leaving this section, it is important to emphasize that both the present work and simulations by Prakash et al. [1] used turbulence models of the same level (two-equation closure) and that both sets of results reproduced the main features of the experimental data. As already mentioned, in [1] arguments were presented for justifying the use of a k-ε model for the present configuration, in spite of the well-established shortcomings of such model. In this work, the same view expressed in Prakash et al. [1] is shared by the authors.

Table 3.1 Porous medium properties for simulated foams (taken form Prakash et al. [1, 5])

Foam	G10	G30	G45	G60
ϕ	0.971	0.9755	0.978	0.976
K (m^2)	2.84E-7	6.95E-8	1.6E-8	1.2E-8
c_F (m)	0.1227	0.1218	0.1232	0.1161

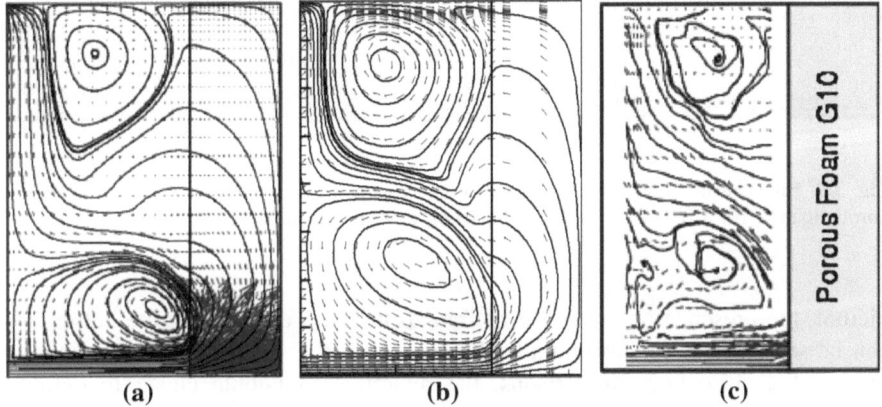

(a)   (b)   (c)

Fig. 3.8 Comparison of streamfunction pattern for $H = 0.10$ m, $h_p = 0.05$ m and porous foam G10: **a** CFD results—Prakash et al. [1, 5], **b** present results, **c** LDV measurements—Prakash et al. [1, 5]

3.2   Porous Medium

As mentioned, two different thicknesses h_p were considered for the porous layer at the bottom of the cylinder, namely $h_p = 0.10$ and 0.05 m. Porous medium properties are given in Table 3.1, where the values are the same as those for metallic porous foams used by Prakash et al. [1, 5]. All porosities considered are nearly equal and close to unity, so that the changes on the porous material effectively represent a change in its permeability. Material G10 is the most permeable. According to the nomenclature used in Prakash et al. [1, 5], the label for the metallic foams is associated with its number of pores per inch, ppi. Thus, material G10 had 10 ppi.

3.2.1   Mean Flow

Figure 3.8 shows a comparison of the numerically simulated streamfunctions with experimental flow visualizations and CFD data from Prakash et al. [1, 5] for $H = 0.10$ m, $h_p = 0.05$m and $Re = 30,000$ for the porous foam G10. From the

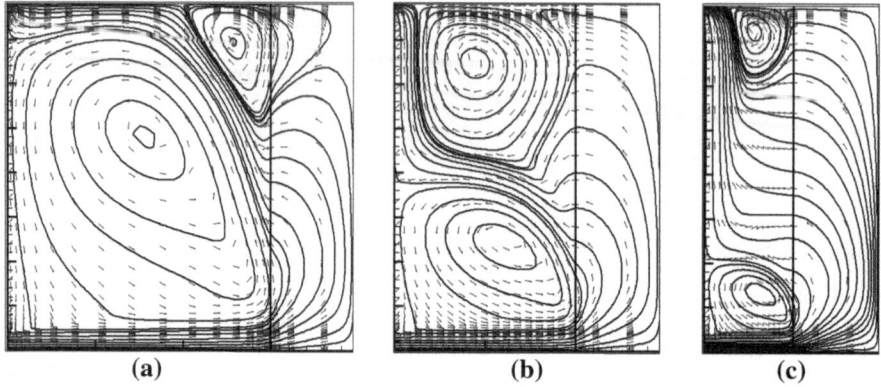

Fig. 3.9 Effect of the fluid layer height on stream function for $h_p = 0.05$ m, $Re = 30,000$ and porous foam G10: **a** $H = 0.15$ m, **b** $H = 0.10$ m, **c** $H = 0.05$ m

picture, the presence of two large recirculation zones dominating the entire flow can be seen, a feature substantially different from the single recirculation shown before (Fig. 3.1). For comparisons, the recirculating bubble closer to cylinder centerline will be called *primary* whereas the one near the cylinder wall will be named *secondary* recirculation. For the present case, both primary and secondary recirculations seem to be slightly larger than those from flow visualizations with LDV, but in general, results seem to be in good agreement with experimental data. The streamlines inside the porous medium cannot be seen by flow visualization methods due to the opaque characteristic of the porous material. They however can be predicted from the flow behavior in the fluid layer above the foam.

The effect of the height of the fluid layer is shown next in Fig. 3.9, showing the streamfunction behavior for $H = 0.15$, 0.10 and 0.05 m. For $H = 0.15$ m, the primary recirculation is much stronger than the secondary one, indicating a smaller effect of the porous layer on the flow structure in the free flow region. Under such conditions, the flow behavior tends to be similar to that obtained in a cylinder without the porous layer. This happens due to the spread of the jet before colliding into the porous foam. With a decrease in H, the primary recirculation tends to decrease in size as the secondary one grows, so that for $H = 0.10$ m both bubbles have almost the same dimensions, filling almost completely the fluid layer and compressing the streamlines between both recirculations. For $H = 0.05$ m, both recirculations decrease in size, moving their centers apart, towards the cylinder centerline (*primary*) and in the direction of the cylinder wall (*secondary*). Streamlines between both recirculations are distributed rather regularly, with the flow nearly perpendicular to the interface as fluid leaves the porous foam to the fluid layer, a different behavior from that occurring for higher values of H.

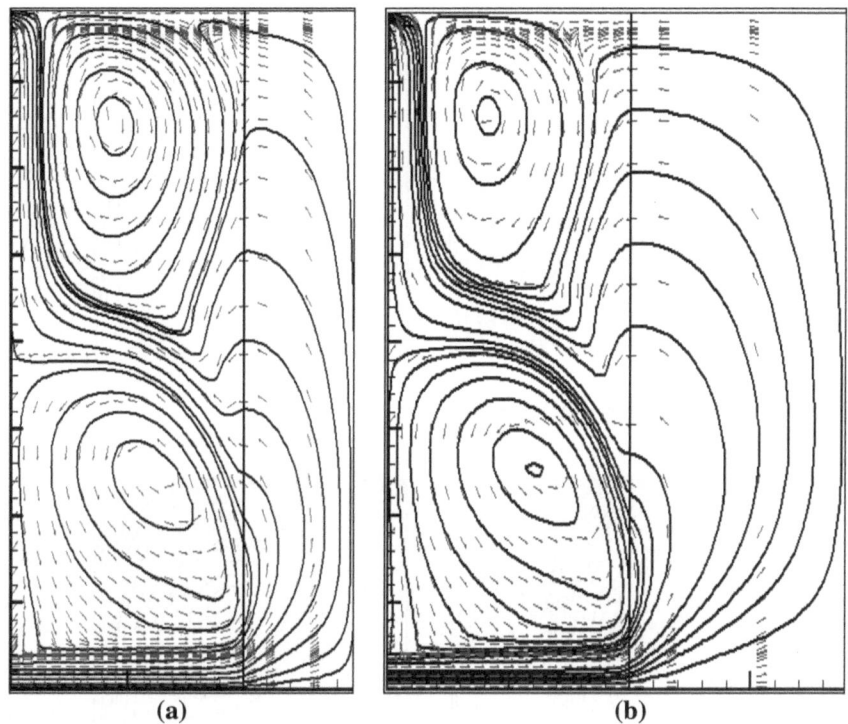

Fig. 3.10 Effect of porous layer thickness on streamfunction for $H = 0.10$ m, $Re = 30,000$ and porous foam G10: **a** $h_p = 0.05$ m, **b** $h_p = 0.10$ m

The effect of the porous layer thickness, h_p, is presented in Fig. 3.10 while maintaining the same fluid layer height, H. From the picture, it can be inferred that the thickness of the porous foam has less influence on the main flow behavior in comparison with the effect caused by varying H (see Fig. 3.9) With an increase in h_p, the primary recirculation shows only a small increase in size while the secondary one decreases a little. No other significant effect can be detected.

The effect of the porous layer material, effectively representing a change in its permeability, is shown in Fig. 3.11. For the porous foam G10 (highest permeability), a secondary recirculation develops with considerable size close to the cylinder wall. For the other foams G30, G45 and G60, this recirculation decreases due to the reduction of the porous layer permeability, so that the porous layer tends to act as a solid obstacle being hit by a jet, as in the previous cases presented in Figs. 3.1, 3.2 and 3.4.

Figure 3.12 show radial and axial position of central recirculation (close to the symmetry axis) for porous foam G10 with $h_p = 0.05$ m. From the pictures, a pattern can be distinguished, where the radial position (Fig. 3.12a) of the recirculation tends to move in the direction of the cylinder wall for an increase in H.

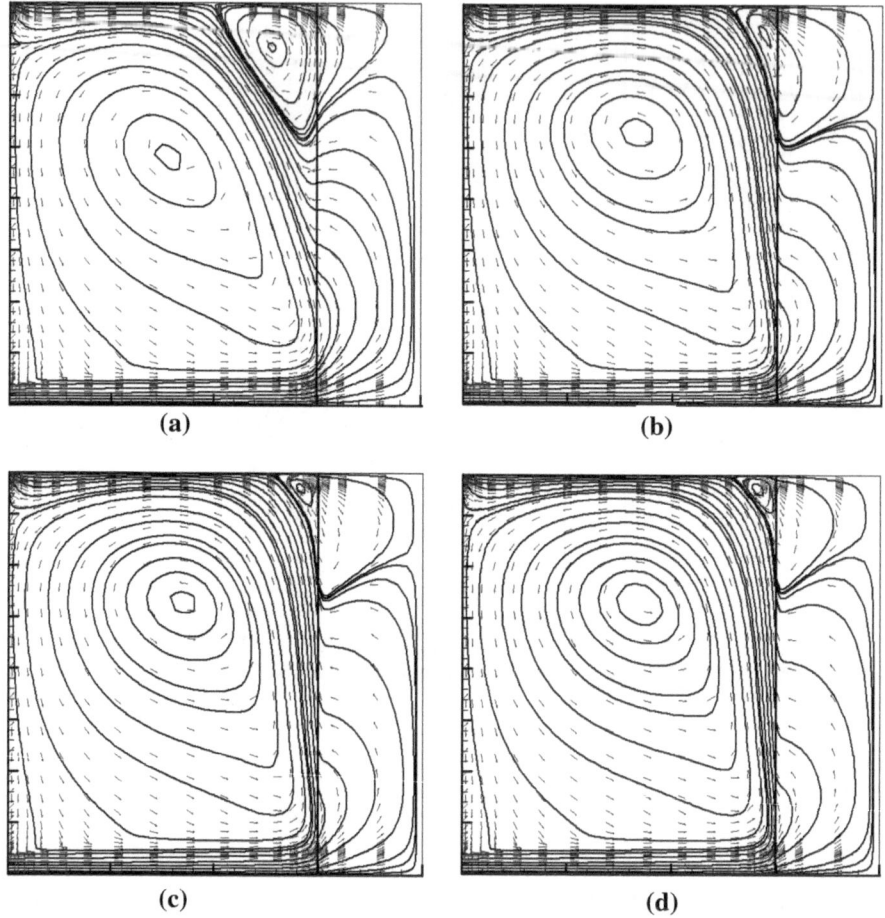

Fig. 3.11 Effect of porous medium material on streamfunction for $H = 0.15$ m, $h_p = 0.05$ m (see properties in Table): **a** G10, **b** G30, **c** G45, **d** G60

Although the results are reasonably good for smaller values of H, simulation behavior for the higher values of H does not satisfactorily represent the central recirculation position along the radial axis. For the axial position (Fig. 3.12b), another pattern can be noticed. While the LDV measurements indicates a movement of the circulation center in the direction to the collision plate, as H increases, the present simulations show movement in the opposite direction. Experimental measurements and present results shows a very good agreement for $H = 0.1$, but rather poor agreement for $H = 0.05$ and 0.15. This could be due to mesh sensitivity, with the mesh for $H = 0.1$ representing an optimal grid point distribution for that specific case. In fact, optimal grid distribution was investigated only for the case $H = 0.1$ whereas for $H = 0.05$ and 0.15 the same grid layout was applied.

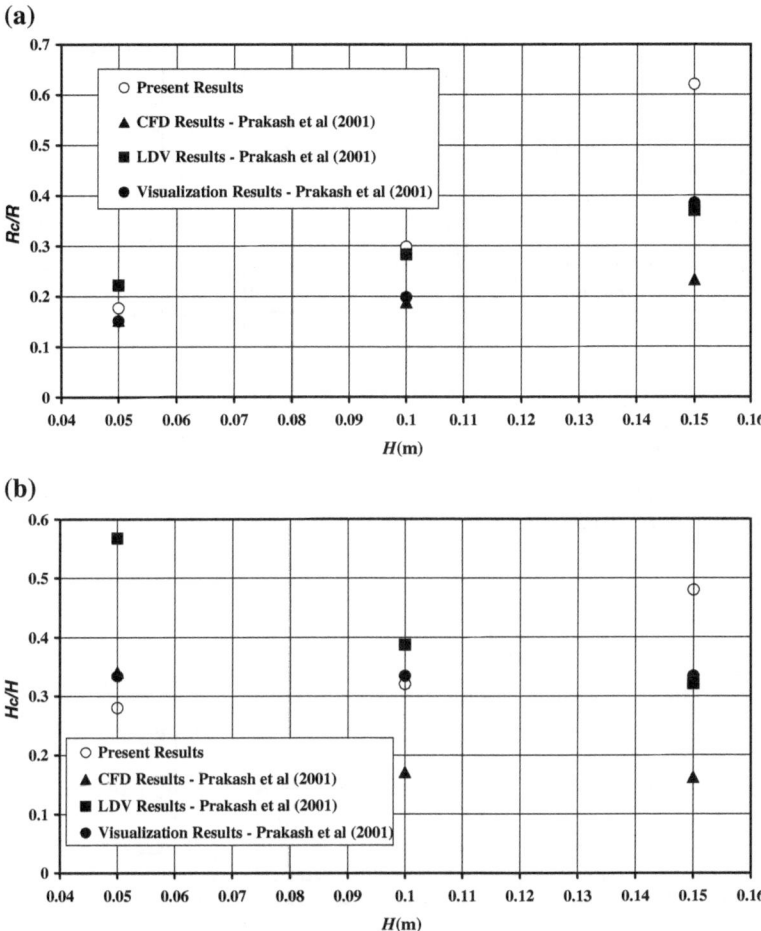

Fig. 3.12 Position of center of primary recirculation as a function of the height of fluid layer for porous foam G10 and $h_p = 0.05$ m: **a** radial position, **b** axial position

Results for the axial velocity profiles are shown in Fig. 3.13, for porous foam G10 and $h_p = 0.05$ m and $Re = 30{,}000$. For all three fluid layers, it can be seen that both numerical simulations seems to slightly overpredict the actual flow behavior close to the jet centerline. For $H = 0.15$ m, numerical predictions are closer to experimental data in comparisons with $H = 0.10$ and 0.05 m. The general over prediction of the velocities profiles might be due to the fact that, in the experiments, the flow at the jet exit was not fully developed, a condition which was assumed in the present simulations. Figure 3.14 presents the radial profiles for the same parameters as in the previous figure. For higher values of z/H, simulations seem to present difficulties in following experimental data, even though, close to

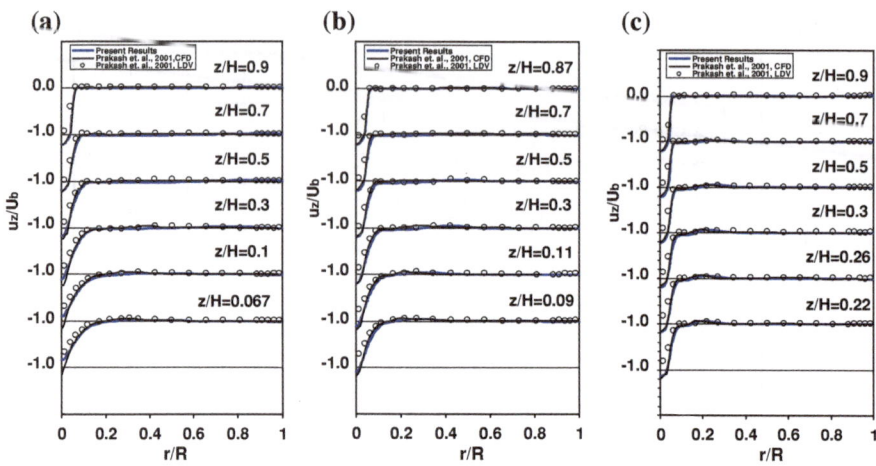

Fig. 3.13 Axial velocity profiles for $h_p = 0.05$ m on porous foam G10: **a** $H = 0.15$ m, **b** $H = 0.10$ m, **c** $H = 0.05$ m

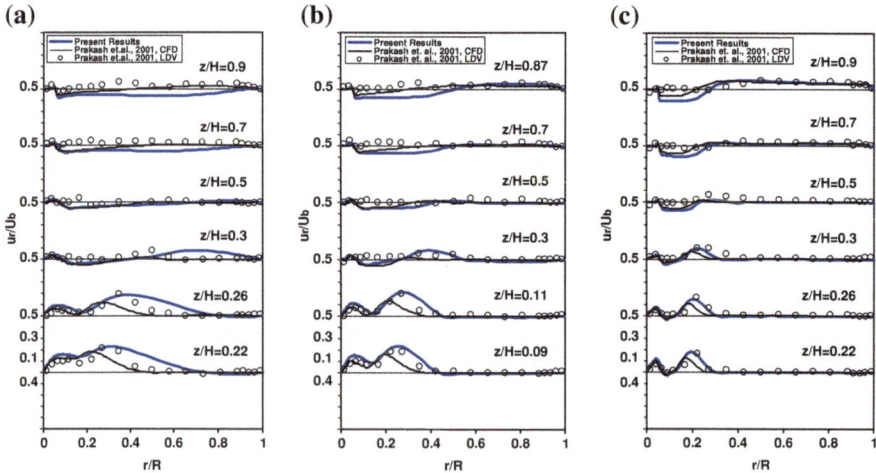

Fig. 3.14 Radial velocity profiles for $h_p = 0.05$ m on porous foam G10: **a** $H = 0.15$ m, **b** $H = 0.10$ m, **c** $H = 0.05$ m

the interface region, the flow behavior is reasonably well predicted, particularly for $H = 0.10$ m. For the second hump along radial profiles, while numerical simulations by Prakash et al. [1, 5] under predict experimental values, the present simulations over predicts them by a small amount.

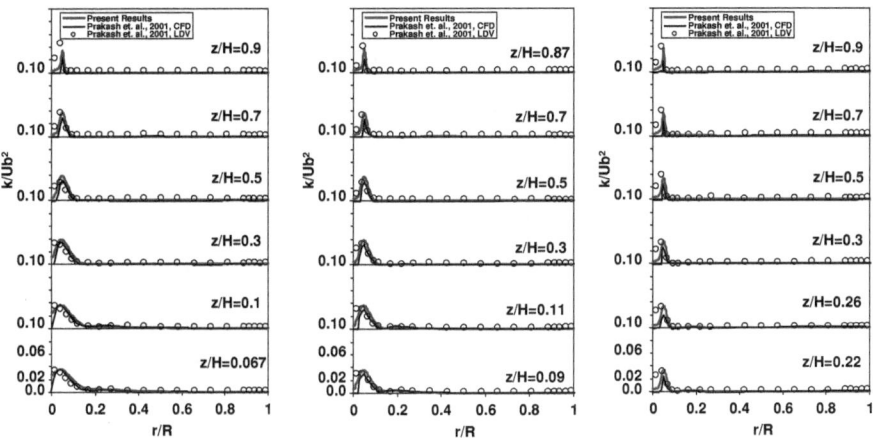

Fig. 3.15 Turbulence kinetic energy profiles for $h_p = 0.05$ m on porous foam G10: **a** $H = 0.15$ m, **b** $H = 0.10$ m, **c** $H = 0.05$ m

3.2.2 Turbulent Field

Turbulence kinetic energy profiles are shown in Fig. 3.15. As in the case of clear medium (Fig. 3.7), the present results give a good prediction for the profiles, following closely the peak value of turbulent energy close to the cylinder central axis. Finally, Fig. 3.16 shows turbulent kinetic energy contours from Graminho and de Lemos [7] for $h_p = 0.05$ m, porous foam G10 and $Re = 30,000$. From the picture, it can be seen that turbulence penetrates into the porous medium, as can be noticed by the contour lines that goes inside the porous bed. As the jet penetrates the foam, calculated turbulence intensities with the present model seem to be lower than those by Prakash et al. [1, 5], who used only the Darcy term (dashed lines in Fig. 3.16b, d, f). However, calculated levels of turbulence herein are higher than those computed by the same authors using both the Darcy and Forchheimer terms (solid lines). This second set of results by Prakash et al. [1, 5] indicates that turbulence is damped almost completely at the interface. At the fluid layer, the present results are in agreement with published data.

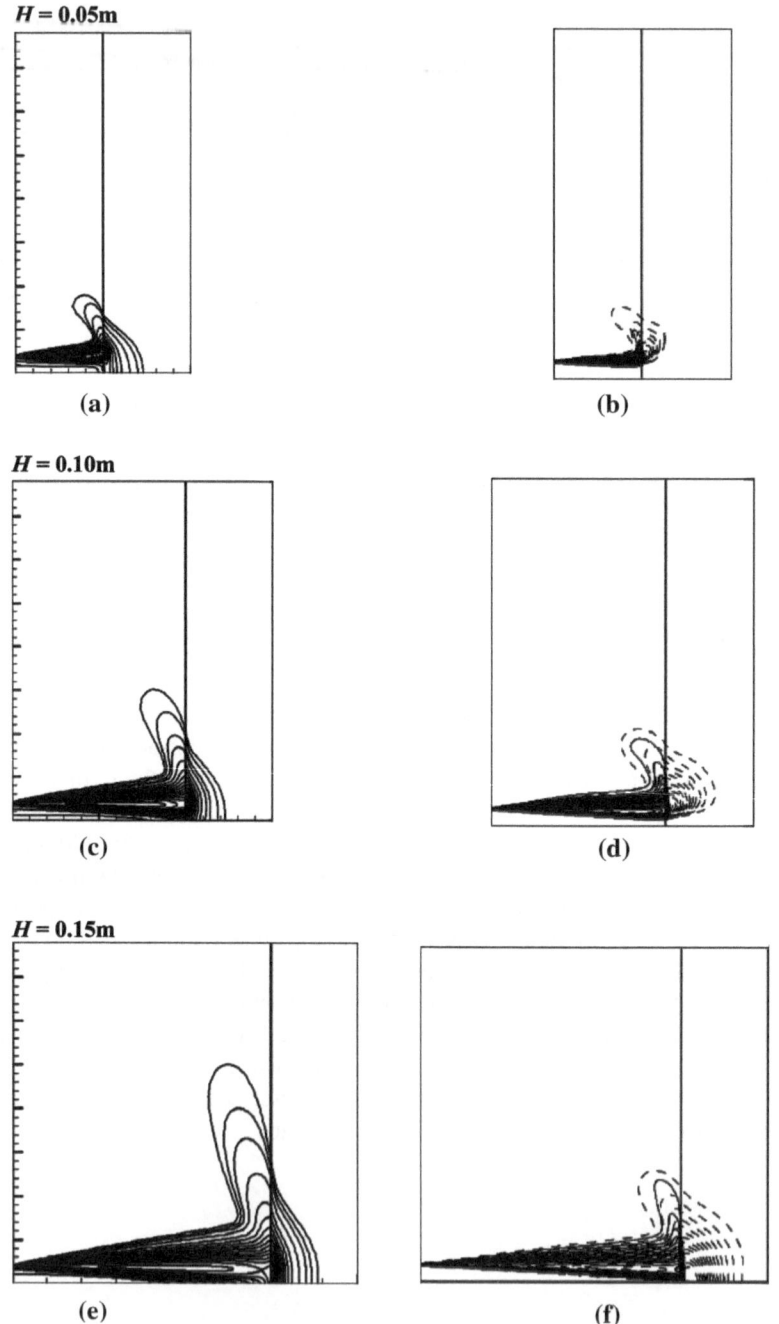

Fig. 3.16 Comparison of turbulence kinetic energy contours for $h_p = 0.05$ m on porous foam G10: *Left* **a,c,e** Graminho and de Lemos [7], *Right* **b,d,f** CFD simulations—Prakash et al. [1, 5] where: *Dashed lines*—Darcy term only, *Solid lines*—Darcy and Forchheimer terms

References

1. M. Prakash, F.O. Turan, Y. Li, J. Manhoney, G.R. Thorpe, Impinging round jet studies in a cilindrical enclosure with and without a porous layer: Part I: Flow visualizations and simulations. Chem. Eng. Sci. **56**, 3855–3878 (2001)
2. B.V. Antohe, J.L. Lage, A general two-equation macroscopic turbulence model for incompressible flow in porous media. Int. J. Heat Mass Transf. **40**, 3013 (1997)
3. M.H.J. Pedras, M.J.S. de Lemos, Macroscopic turbulence modeling for incompressible flow through undeformable porous media. Int. J. Heat Mass Transf. **44**(6), 1081–1093 (2001)
4. M.J.S. de Lemos, Fundamentals of the double—decomposition concept for turbulent transport in permeable media. Materialwissenschaft und Werkstofftechnik **36**(10), 586–593 (2005)
5. M. Prakash, F.O. Turan, Y. Li, J. Manhoney, G.R. Thorpe, Impinging round jet studies in a cylindrical enclosure with and without a porous layer: Part II: LDV measurements and simulations. Chem. Eng. Sci. **56**, 3879–3892 (2001)
6. F.P. Incropera, D.P. DeWitt, *Fundamentals of Heat and Mass Transfer* (Wiley, New York, 1990)
7. D.R. Graminho, M.J.S. de Lemos, Simulation of turbulent impinging jet into a cylindrical chamber with and without a porous layer at the bottom. Int. J. Heat Mass Transf. **52**, 680–693 (2009)

Chapter 4
Heat Transfer Using the Local Thermal Equilibrium Model

4.1 Input Parameters for the LTE Model

For an impinging jet, the flow is considered to be turbulent for $Re > 1,000$, where the Reynolds number is given by

$$Re = \frac{\rho \, v_0 \, D_h}{\mu} \tag{4.1}$$

where v_0 is the incoming jet velocity and $D_h = B$ when calculating Re for adequate comparisons with similar simulations in the literature (see Fig. 1.2).

The low Re turbulence model presented above was used is all simulation to follow. In order to guarantee that grids nodes be positioned within the laminar sub-layer, the closest grid node to the wall had a value for its wall coordinate y^+ less than unity ($y^+ \leq 1$). Further, inlet value for the turbulent kinetic energy k at the jet entrance was estimated using:

$$k_0 = \frac{3}{2}(v_0 \, I)^2 \tag{4.2}$$

where I is the turbulence intensity assumed to prevail in the incoming flow. For the dissipation rate of k, ε, the inlet value was calculated according to:

$$\varepsilon_0 = c_\mu^{3/4} \frac{k^{3/2}}{\ell} \tag{4.3}$$

where ℓ is a length scale associated with the energy containing eddies. Table 4.1 summarizes the parameters used as input.

M. J. S. de Lemos, *Turbulent Impinging Jets into Porous Materials*,
SpringerBriefs in Computational Mechanics,
DOI: 10.1007/978-3-642-28276-8_4, © The Author(s) 2012

Table 4.1 Input parameters for turbulent flow simulations with the LTE Model

Fluid	Density ρ	Viscosity μ	B	L	T_0	T_1	Length scale, l	Turbulence intensity, I
Air	1.225 kg/m3	1.789×10^{-5} N.s/m^2	14.23 mm	500 mm	309.1 K	347.6 K	0.07B	2%

4.2 Grid Independence Studies for the LTE Model

Grid validation was conducted with the conditions $Re = 10,400$ and $H/B = 2.6$. At the jet entrance, values in Table 4.1 were employed. For grid independence studies the overall heat power at the impinged wall, given by,

$$Q_w = \int_{x=o}^{x=L} -q''|_{y=H}\, w\, dx, \quad q''|_{y=H} = -k_f \frac{\partial \langle \bar{T} \rangle^i}{\partial y}\Bigg)_{y=H}, \tag{4.4}$$

where $w = 1$ m is the transverse plate width, was calculated for several grids and compared in Table 4.2.

Also, local Nu number along that wall was evaluated by:

$$Nu = \frac{hH}{k_f} = \left(\frac{\partial \langle \bar{T} \rangle^i}{\partial y}\right)_{y=H} \frac{H}{T_1 - T_0} \tag{4.5}$$

and h is a film coefficient. Figure 4.1a shows local Nu distribution calculated according to Eq. (4.5), also as a function of grid size. One can note in the table that for grids greater than 80×216, the deviation in relation to the finest grid is less than 0.5%. As such, all simulation for turbulent flow herein were carried out on a grid of size 80×216 (Fig. 4.1a), which was refined close to the wall and about the jet entrance, where the steepest temperature gradients are expected to occur.

4.3 Clear Channel

The first set of results is related to the configuration shown in Fig. 1.2a, where no porous material is attached to the bottom wall. Once an appropriate grid was chosen, code validation was carried out by comparing Nu numbers calculated at the bottom wall compared with results by Wang and Mujumdar [1], for two cases, namely for $H/B = 6$ and $Re = 5,200$ (Fig. 4.1b) and for $H/B = 2.6$ and $Re = 10,400$ (Fig. 4.1c). The figure indicates that for $H/B = 6$ a good agreement is obtained whereas for $H/B = 2.6$ results do not match quite well experimental values. This might be due to the fact that such flow entails a high degree of complexity, particularly for turbulent flow regime, as discussed by Wang and Mujumdar [1] and Heyerichs and Pollard [2]. Nevertheless, as the main purpose of

Table 4.2 Influence of grid size on integral wall heat flux—LTE Model

Grid size	40×216	80×216	80×400	100×400
Wall heat power Q_w, Eq. (4.4)	772.75 W	818.87 W	825.68 W	820.05 W
Deviation in relation to grid 100×400	6.12%	0.14%	0.68%	0.00%

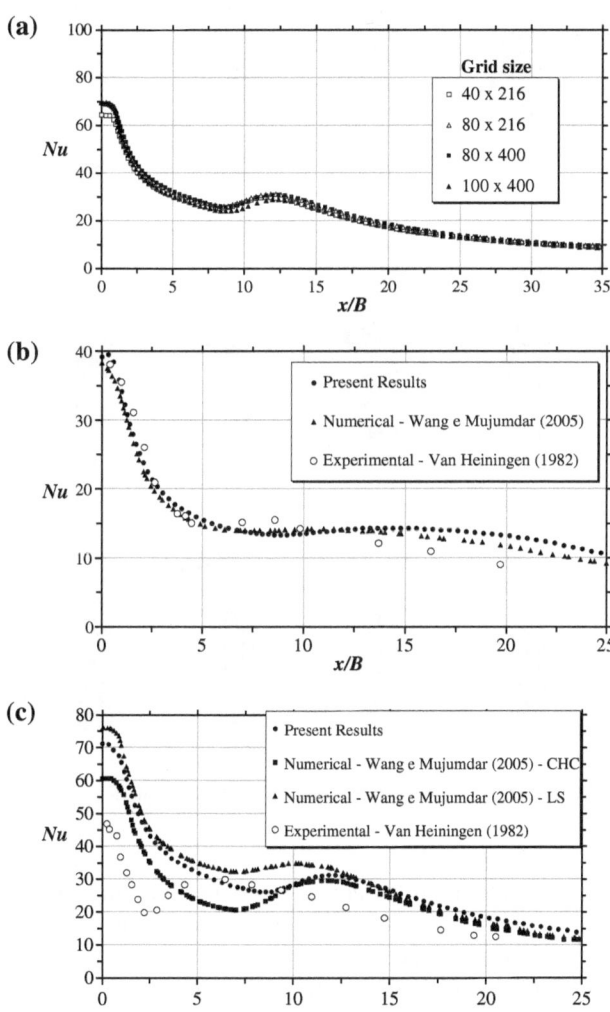

Fig. 4.1 Validation for distribution of Nu along the lower plate for clear channel. **a** Effect of grid size. **b** $Re = 5{,}200$, $H/B = 6$, exp. by Van Heiningen (1982) reported by Heyerichs and Pollard [2]. **c** $Re = 10{,}400$, $H/B = 2.6$, Low Re models by Chang et al. (CHS) and Launder and Sharma (LS) reported by Wang and Mujumdar [1]

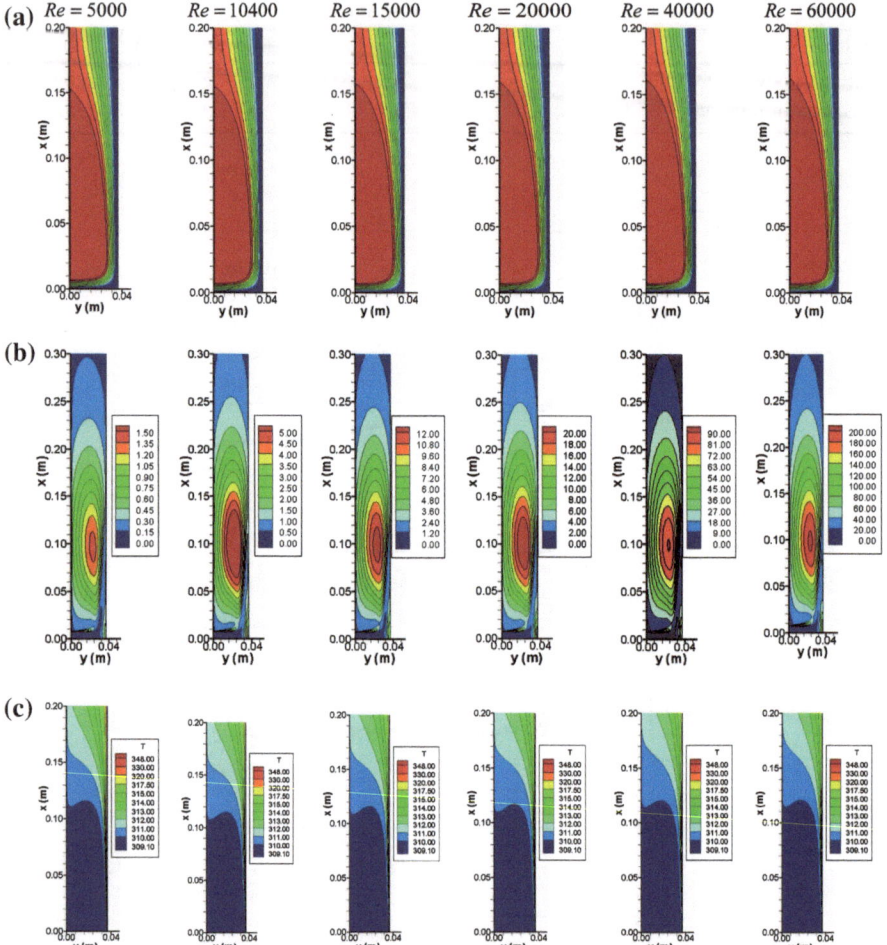

Fig. 4.2 Effect of *Re* for $H/B = 2.6$. **a** Streamlines. **b** Turbulent kinetic energy k. **c** Temperature T

this work is to investigate the influence of a porous layer on heat transfer, and not the turbulence model employed, and considering further the fact that a reasonable agreement with experimental data was achieved, the computer code and the grid size were assumed to be sufficiently accurate for the investigation here conducted.

Figure 4.2a shows streamlines for $H/B = 2.6$ as a function of *Re*, where one can note that the flow pattern is not substantially affected by *Re*, indicating that the fully turbulent regime is achieved. Further, the flow is characterized by a large recirculation zone attached to the jet entrance, named primary vortex. Corresponding statistical field is presented in Fig. 4.2b. In the figure one can note similarity among the maps and an increase on turbulence intensity as *Re* increases. Next, the temperature filed is presented in Fig. 4.2c. Likewise the hydrodynamic

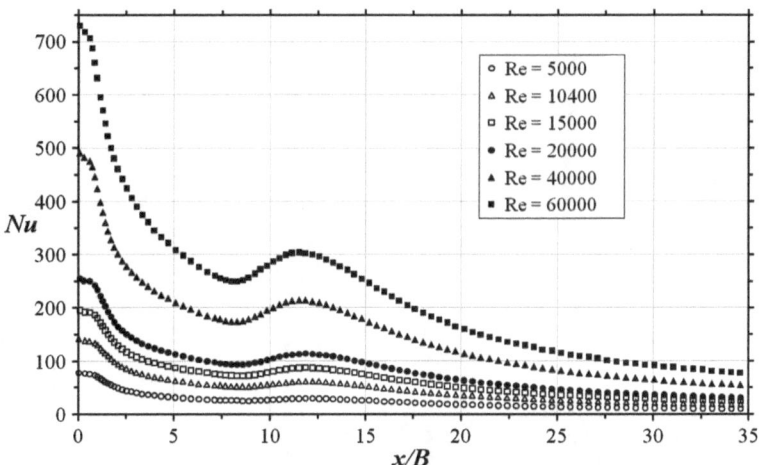

Fig. 4.3 Effect of *Re* on *Nu* number for clear channel

field, no substantial changes are detected as *Re* increases. It is also observed that isolines are close to each other in the stagnation region, which characterizes a thin thermal boundary layer with high temperature gradients in that region.

For a performance analysis on heat transfer due to the jet velocity variation, Fig. 4.3 shows the local *Nu* close to the target plate for various Reynolds numbers. According to Fig. 4.3, as the Reynolds number increases, the curves for Nusselt shift towards higher values increasing the peak at the stagnation region ($x = 0$). In addition, it can be seen that a second Nusselt peak appears for $Re > 10,400$ and at about $x/B = 12$, becoming more pronounced as *Re* increases [3].

4.4 Channel with Porous Layer

When a layer of porous material is added to the bottom of the channel, the resulting configuration is shown in Fig. 1.2b. The material is assumed to be rigid, with porosity ϕ, non-dimensional thickness h/H, Darcy number $Da = K/H^2$ and thermal conductivity ratio k_s/k_f. Results below are obtained using distinct values for such four parameters.

4.4.1 Effect of Porosity, ϕ

In this section, the results were obtained using $H/B = 2.6$, $k_s/k_f = 10$, $Re = 10,400$, $Da = 8.95 \times 10^{-5}$ and $h/H = 0.50$.

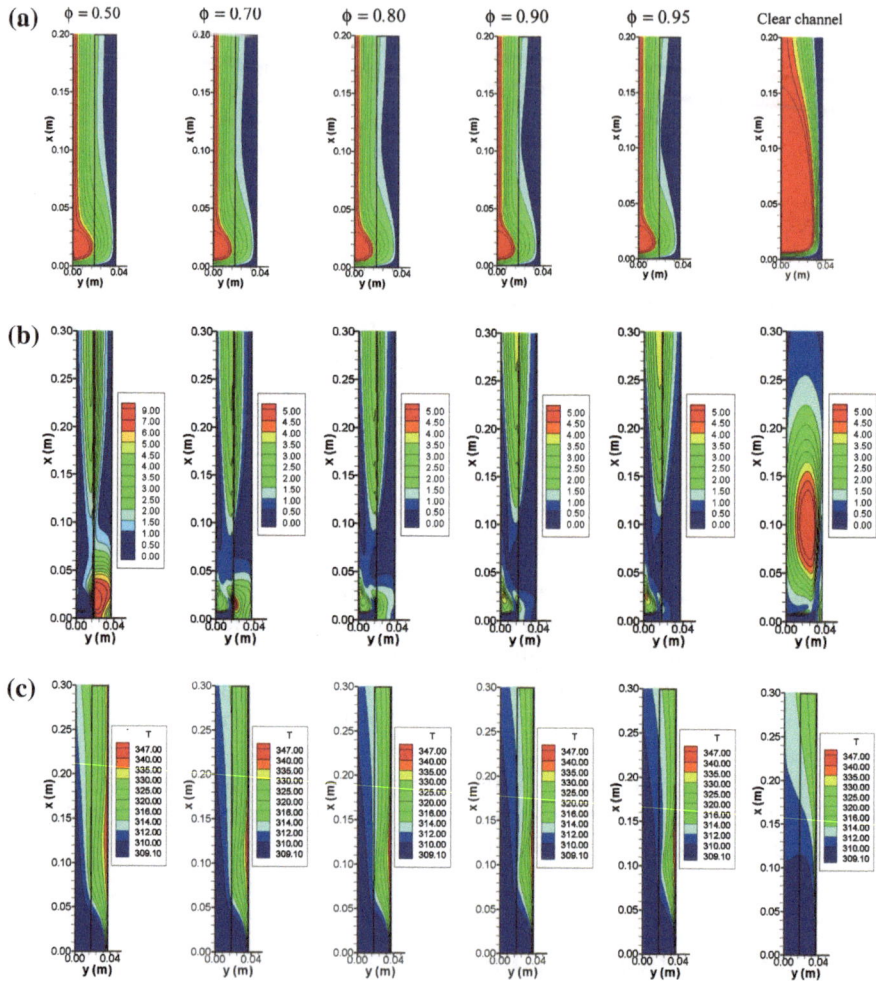

Fig. 4.4 Effect of porosity ϕ for $H/B = 2.6$, $k_s/k_f = 10$, $Re = 10,400$, $Da = 8.95 \times 10^{-5}$ and $h/H = 0.50$. **a** Streamlines. **b** Turbulent field k. **c** Temperature field T

Figure 4.4a shows streamlines and indicates that porosity variation does not strongly influences the flow behavior, as also confirmed by Graminho and de Lemos [4] and de Lemos and Fischer [3].

One can note that the presence of the porous layer reduces de size of the primary vortex and that the strength of convection fluxes is smaller inside the porous material than in the clear passage, as expected. After the stagnation region, in the accelerating length ($0.06 < x < 0.10$), the flow tends to detach from the porous layer, reattaching further downstream. This trend is more pronounced as porosity increases, which is expected since within low porosity media the flow

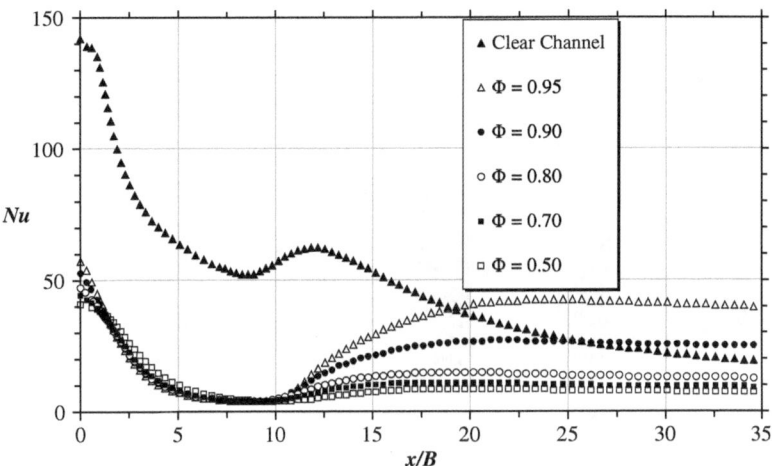

Fig. 4.5 Local Nusselt distribution for various porosities with $H/B = 2.6$, $k_s/k_f = 10$, $Re = 10,400$, $Da = 8.95 \times 10^{-5}$ and $h/H = 0.50$

tends towards a plug-flow configuration, as it appears for be the case with $\phi = 0.50$.

Figure 4.4b shows corresponding results for the turbulent field. One note that as porosity decreases, k levels increase. High values of k are also encountered around the jet entrance where steep velocity gradients occur. Around the interface, levels of k are also high. This scenario contrasts with the clear channel distribution where most of the turbulence energy is generated in the recirculation zone corresponding to the primary vortex.

Porosity effects on T are presented in Fig. 4.4c. In the figure one can see that isolines present a coherent behavior with the flow pattern and, in the region $0.06 < x < 0.10$, such lines bulge away from the bottom wall. Further, it becomes evident that for low porosities the temperature is more homogenized and the thermal boundary layer becomes thicker, resulting in lower temperature gradients at the wall.

Nusselt numbers at wall are presented next in Figure 4.5, showing a peak in the stagnation region and a minimum value in the range $x/B = 5.0 - 10.0$, which is related to the above seen flow structure within $0.06 < x < 0.10$, where velocities are low and boundary layers thick. Also, one can observe that porosity does not affect Nu in the stagnation region, but increases the Nusselt number for a higher ϕ as the flow downstream resembles a wall jet. For low porosities, the value for Nu downstream the flow is reduced and a minimum value for its longitudinal distribution tends to disappear, indicating that under such circumstances the temperature is more uniform, as already seem in Fig. 4.4.

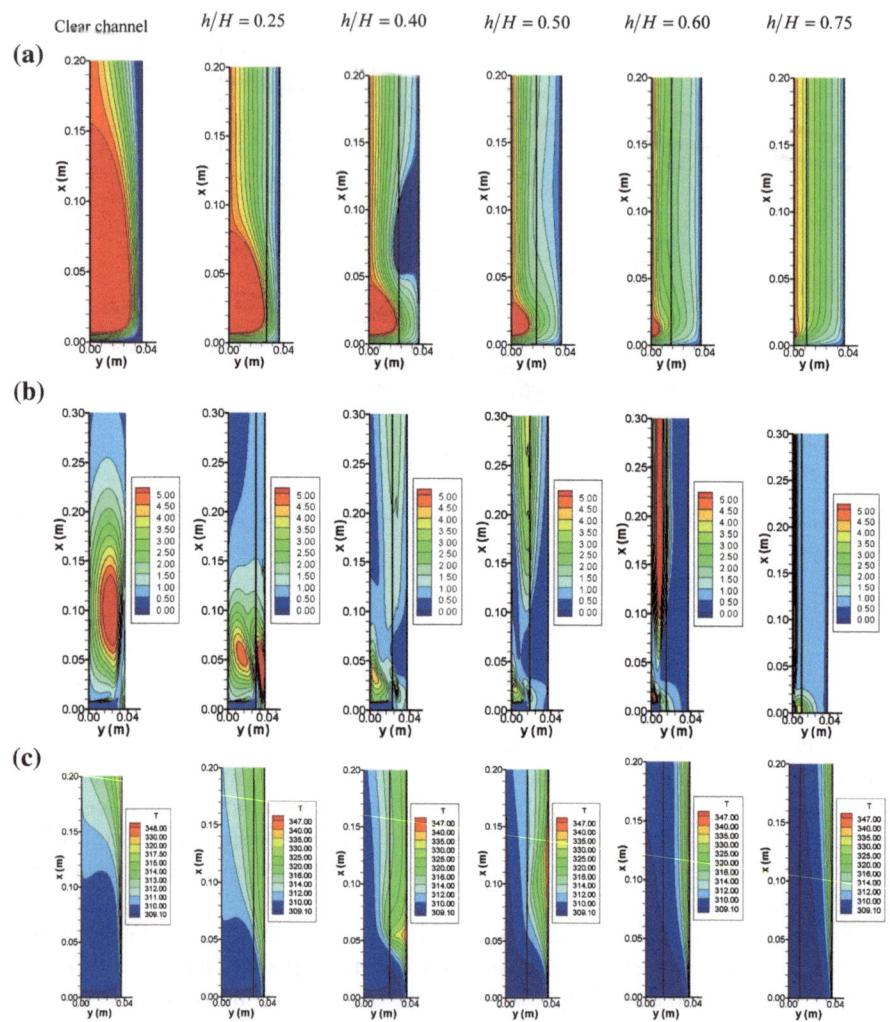

Fig. 4.6 Effect of blockage ratio h/H for $H/B = 2.6$, $k_s/k_f = 10$, $Re = 10,400$, $Da = 8.95 \times 10^{-5}$ and $\phi = 0,90$. **a** Streamlines. **b** Turbulent field k. **c** Temperature field T

4.4.2 Effect of Channel Blockage, h/H

A study of the influence of the porous layer thickness in the heat transfer is now presented. The streamlines for a simulation with various porous layer thicknesses, with $H/B = 2.6$, $k_s/k_f = 10$, $Re = 10,400$, $Da = 8.95 \times 10^{-5}$ and $\phi = 0.90$, are presented in Fig. 4.6a. The figure indicates that the value of h/H has a great influence on the flow pattern. As the blockage ratio increases, the primary vortex is reduced, being nearly extinguished for $h/H = 0.75$. Specifically for $h/H = 0.40$, a

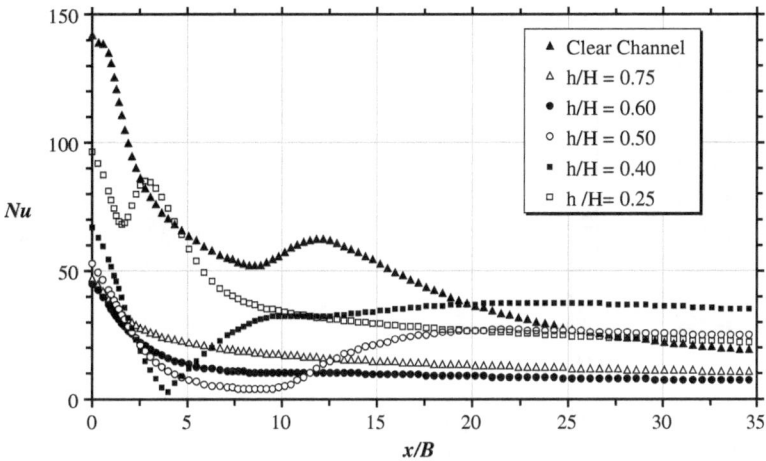

Fig. 4.7 Local Nusselt distribution for various blockage ratios h/H with $H/B = 2.6$, $k_s/k_f = 10$, $Re = 10,400$, $Da = 8.95 \times 10^{-5}$ and $\phi = 0.90$

new recirculation region seems to appear close to the bottom wall, whose vortex is here named secondary. Also, as h/H increases, streamlines tends to become more uniform within the channel.

The corresponding statistical field behavior is presented in Fig. 4.6b, where maps for k are shown. For h/H from 0.25 to 0.50, levels of k are reduced as the porous layer gets thicker. On the other hand for h/H higher than 0.60, steep velocity gradients in the fluid layer increases k since the fluid is pushed towards the free gap, causing a large velocity gradients in that regions. In addition, the temperature field in Fig. 4.6c is also influenced by the porous layer thickness variation. The thermal field becomes more homogenized for thicker layers, as a consequence of the more uniform flow field observed in Fig. 4.6a.

To complete the analysis the local Nusselt number is presented in Fig. 4.7 for various porous layer thicknesses. The stagnation Nusselt peak diminishes with the insertion of the porous layer, and for thicknesses less than, $h = 0.40H$, the second peak in Nu is still present as shown in Fig. 4.7. The presence of the second peak is connected with the secondary vortices appearing in Fig. 4.6. Also, it can be seen that variation of h/H does not influence the value of stagnation Nusselt peak as strongly as the variation of porosity, as can be concluded by comparing Fig. 4.5 and Fig. 4.7. The main influence of the thickness of the porous layer is in shape of the curve of Nusselt distribution along the stagnation region, changing from a double humped profile for clear channel to a single peak distribution for $h/H = 0.75$.

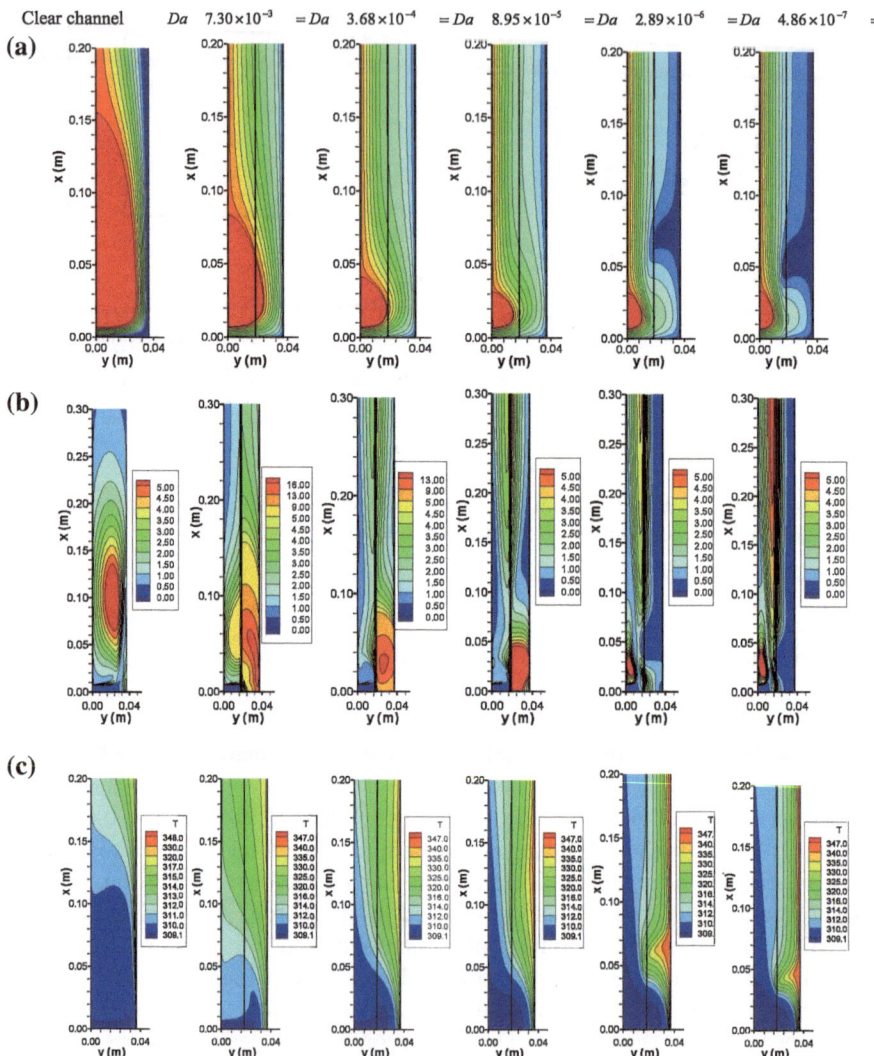

Fig. 4.8 Effect of Da for $H/B = 2.6$, $k_s/k_f = 10$, $Re = 10,400$, $\phi = 0.50$ e $h/H = 0.50$.
a Streamlines. **b** Turbulent field k. **c** Temperature field T

4.4.3 Effect of Darcy Number, Da

Contrary to what was observed in the analysis of the effects of porosity, permeability strongly influences the hydrodynamic field as can be seen in Fig. 4.8a, which was calculated with $H/B = 2.6$, $k_s/k_f = 10$, $Re = 10,400$, $\phi = 0.50$ and $h/H = 0.50$. The size of the primary vortex decreases with a decrease in

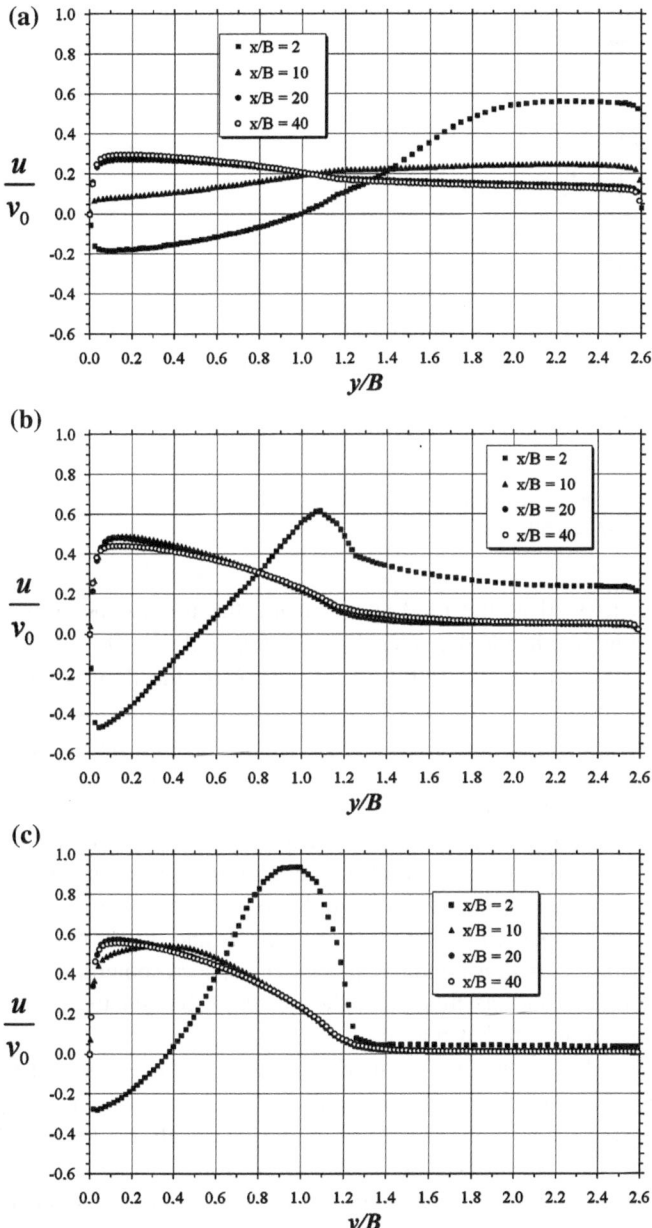

Fig. 4.9 Velocity profiles at several axial stations for $H/B = 2.6$, $k_s/k_f = 10$, $Re = 10,400$, $h/H = 0.50$. **a** $Da = 7.30 \times 10^{-3}$. **b** $Da = 8.95 \times 10^{-5}$. **c** $Da = 4.86 \times 10^{-7}$

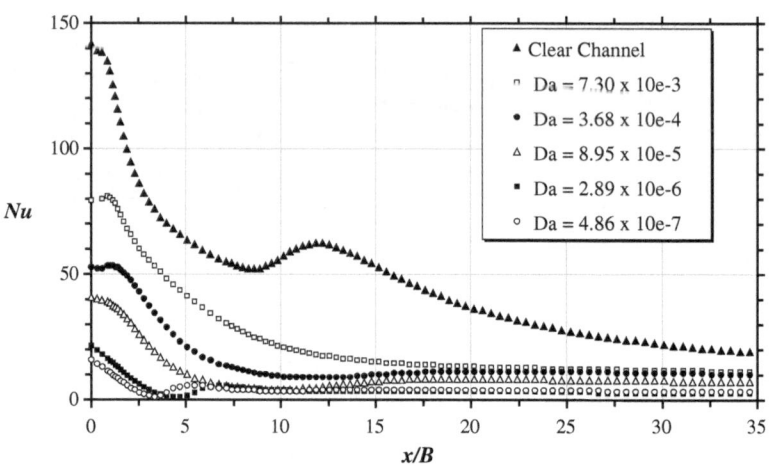

Fig. 4.10 Local Nusselt distribution for various Da with $H/B = 2.6$, $k_s/k_f = 10$, $Re = 10,400$, $\phi = 0.50$ and $h/H = 0.50$

permeability, as well as the fluid speed inside the porous medium leading to a lower mass flow rate inside the layer covering the wall. For $Da = 2.89 \times 10^{-6}$ and $Da = 4.86 \times 10^{-7}$, from $x = 0.05$ m to $x = 0.10$ m, the flow nearly vanishes in the porous material as can be seen by the streamlines in Fig. 4.8a and in the plots shown in Fig. 4.9 for the velocity profiles along the axial channel position.

Back to the analysis of the previous figure, Fig. 4.8b presents two-dimensional fields for the turbulence kinetic energy k. For $Da = 7.30 \times 10^{-3}$ and $Da = 3.68 \times 10^{-4}$, generation of k is enhanced and is even grater than for the clear channel case. Again, the region of highest turbulence corresponds to the jet entrance region where steep velocity gradients produce high generation rates of k. For very low permeabilities, $Da = 4.86 \times 10^{-7}$, and due to the fact that within the porous matrix velocities attain very low values, corresponding k values are also low. On the other hand, as fluid is pushed to flow thought the clear gap, higher values of k are found in that region.

Figure 4.8c, shows temperature distributions for distinct values of $Da = K/H^2$. In this figure one can see that as permeability decreases, isothermal lines, referent to the highest temperatures, bulge from the hot wall thickening the thermal boundary layer at the surface. In accordance with the flow pattern (Fig. 4.8a), cases with $Da = 2.89 \times 10^{-6}$ and $Da = 4.86 \times 10^{-7}$ present temperature peaks from $x = 0.05$ m to $x = 0.10$ m due to flow blockage in that region.

Figure 4.10 shows Nusselt numbers along the wall as a function of Da, where one can see that in such cases Nu at the stagnation region is most influenced, being reduced as Da is reduced. In the wall jet region, local Nu is also reduced, but less intensively when compared with its variation as a function of porosity (Fig. 4.4).

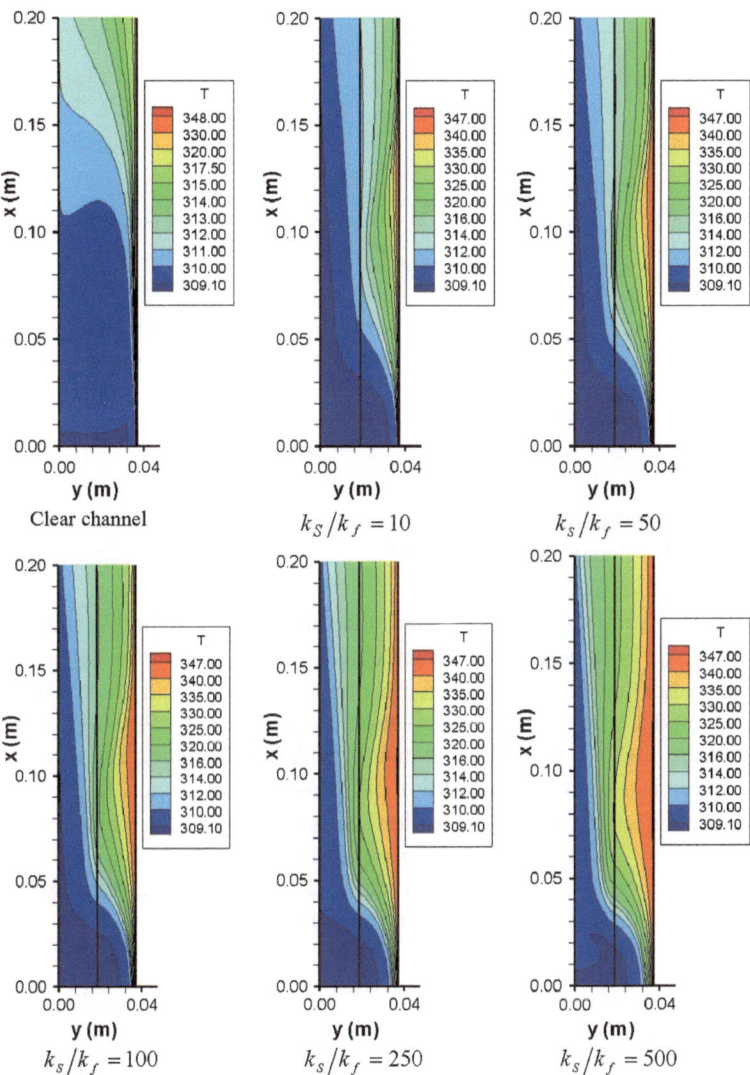

Fig. 4.11 Effect of k_s/k_f on T, $H/B = 2.6$, $Re = 10,400$, $Da = 8.95 \times 10^{-5}$, $\phi = 0.90$ and $h/H = 0.50$

4.4.4 Effect of Thermal Conductivity Ration k_s/k_f

In this section the effect of varying k_s/k_f is investigated. Used parameters are $H/B = 2.6$, $Re = 10,400$, $Da = 8.95 \times 10^{-5}$, $\phi = 0.90$, $h/H = 0.50$ and several values of k_s/k_f. Evidently, since we are running only decoupled solutions, the flow filed is not impacted by the thermal conductivity ratio.

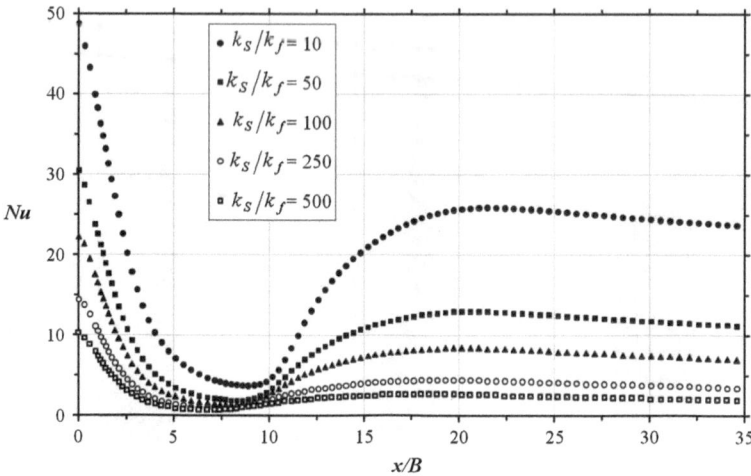

Fig. 4.12 Local Nusselt distribution for various ratios k_s/k_f with $H/B = 2.6$, $Re = 10,400$, $Da = 8.95 \times 10^{-5}$, $\phi = 0.90$ and $h/H = 0.50$

Figure 4.11 shows that, as expected, the thermal field is highly influenced by the ratio k_s/k_f. Fore high thermal conductivity ratios, isotherms depart from the hot wall indicating reduction of temperature gradients and thickening of the thermal boundary layer. Such observation is endorsed by Fig. 4.12 that shows Nusselt at the hot wall. One can note that, in both the stagnation and along the wall regions, there is a substantial reduction of Nu as k_s/k_f increases. It is important to emphasize that tortuosity and dispersion are mechanisms here neglected and that they might play a certain role when computing the temperature field. This is particularly important if one inspects Eq. (2.27) and sees that tortuosity is proportional to the difference between thermal conductivities.

4.5 Integral Wall Heat Flux

As pointed out by de Lemos and Fischer [3], another important parameter to evaluate the effectiveness in using porous layers is to calculate the integrated heat transferred from the bottom wall. Such overall heat transferred from the lower wall to the flowing fluid can be calculated for both configurations presented in Fig. 1.1, as

$$q_w = \frac{1}{L} \int_0^L q_{wx}(x)dx; \quad q_{wx} = -k_{eff} \left. \frac{\partial \langle T \rangle^i}{\partial y} \right|_{y=H},$$

$$k_{eff} = \begin{cases} \phi k_f + (1 - \phi)k_s & \text{with a porous layer} \\ k_f & \text{for clear channel} \end{cases} \tag{4.6}$$

Fig. 4.13 Integral heat flux ratio at the lower wall for varying porosity ϕ, $H/B = 2.6$ and $k_s/k_f = 10$. **a** Effect of Re. **b** Effect of Da. **c** Effect of h/H

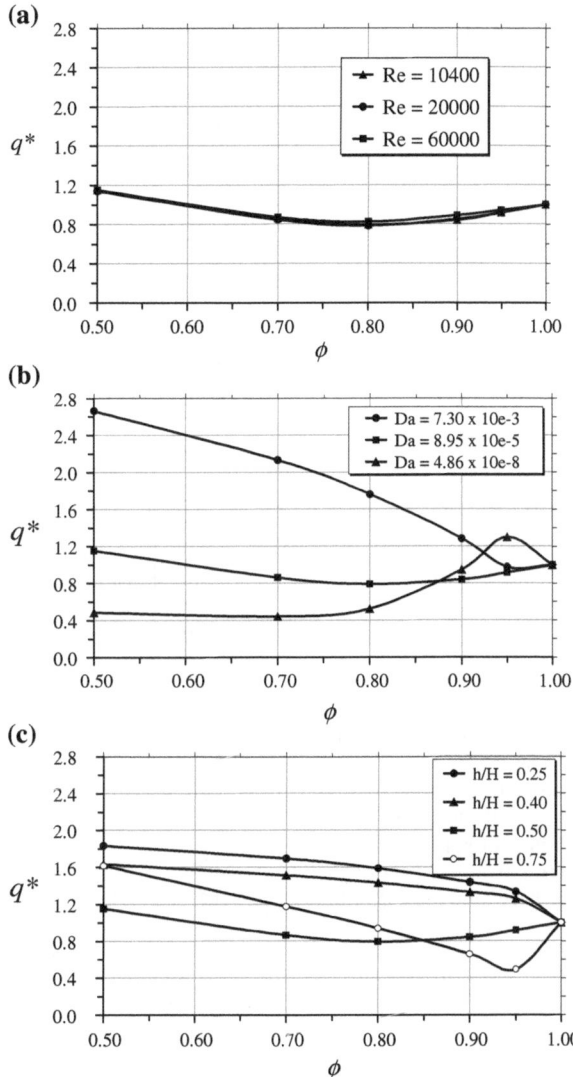

For the cases where the a porous layer is considered, the wall hat flux is given a superscript ϕ on the form q_w^ϕ. The ratio $q^* = q_w^\phi/q_w$ can then be seen as a measure of the effectiveness of using a porous layer for enhancing or damping the amount of heat transferred through the wall.

Figure 4.13 compares the ratio $q^* = q_w^\phi/q_w$ as function of ϕ for several Re, Da and h/H, with $k_s/k_f = 10$. The figures suggests that for cases where there is more solid material per unit volume (low ϕ) and high permeabilities (high Da), there is a net gain when using a porous substrate covering the cooled wall ($q_w^\phi/q_w > 1$). This

Fig. 4.14 Integral heat flux ratio at the lower wall for varying blockage ratio h/H with $H/B = 2.6$, $k_s/k_f = 10$, $Re = 10,400$.
a $Da = 8.95 \times 10^{-5}$.
b $\phi = 0.50$

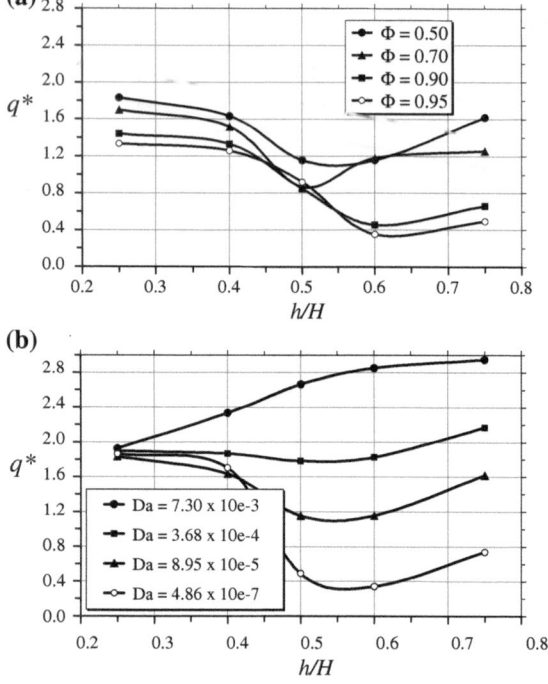

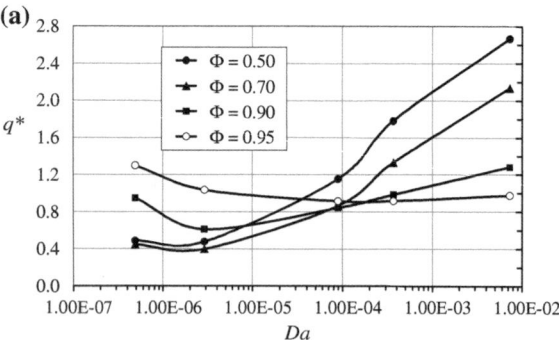

Fig. 4.15 Integral heat flux ratio at the lower wall for various Da, $H/B = 2.6$, $k_s/k_f = 10$, $Re = 10,400$.
a Effect of porosity ϕ, $h/H = 0.50$. **b** Effect of blockage ratio h/H, $\phi = 0.50$

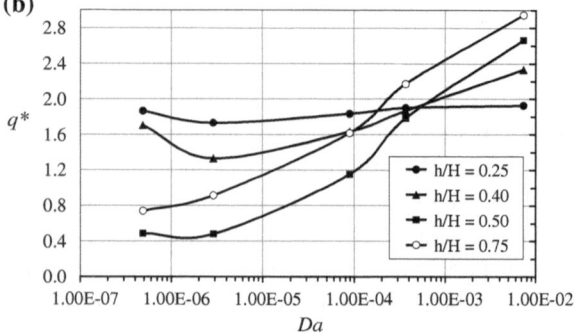

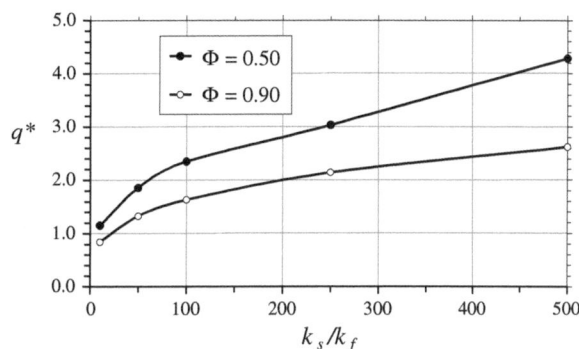

Fig. 4.16 Integral heat flux ratio at the lower wall for various ratios k_s/k_f and porosities with $Re = 10,400$, $h/H = 0.5$, $H/B = 2.6$ and $Da = 8.95 \times 10^{-5}$

is more apparent as the layer gets ticker (Fig. 4.13c). On the other hand, for high porosities ($\phi > 0.6$), and turbulent flow ($Re > 10,000$), disruption of the thermal boundary layer close to the surface, due to the presence of a porous material, damps the overall heat transferred from the wall.

Figure 4.14 presents results for $q^* = q_w^\phi/q_w$ as a function of h/H for distinct ϕ and Da. One note that for low h/H or high Da, the use of a porous material is beneficial to heat transfer. Figure 4.15 further indicates that for less permeable materials, only those with high porosity (Fig. 4.15a) or forming thick layers (Fig. 4.15b) give q_w^ϕ/q_w greater than unity. For more permeable media, a low porosity matrix gives better results, regardless of its thickness.

Finally, Fig. 4.16 compiles results for q_w^ϕ/q_w when the conductivity ratio k_s/k_f is varied. Here, it is important to emphasize that results in the figure were obtained with the LTE or Local Thermal Equilibrium assumption ($\langle T \rangle^i = \langle T_f \rangle^i = \langle T_s \rangle^i$). Such hypotheses might not be valid when the conductivity of the two media differs from each other by a large amount.

Back to Fig. 4.16, one can note that increasing the conductivity ratio past $k_s k_f > 10$, it is always possible to enhance the heat extracted from the bottom wall. The results herein might be useful to design and analysis of energy efficient equipment.

References

1. S.J. Wang, A.S. Mujundar, A comparative study of five low Reynolds number $k - \varepsilon$ models for impingement heat transfer. Appl. Therm. Eng. **25**, 31–44 (2005)
2. K. Heyerichs, A. Pollard, Heat transfer in separated and impinging turbulent flows. Inter. J. Heat Mass Transf. **39**(12), 2385–2400 (1996)
3. M.J.S. de Lemos, C. Fischer, Thermal analysis of an impinging jet on a plate with and without a porous layer. Numer. Heat Transf. A **54**, 1022–1041 (2008)
4. D.R. Graminho, M.J.S. de Lemos, Simulation of turbulent impinging jet into a cylindrical chamber with and without a porous layer at the bottom. Inter. J. Heat Mass Transf. **52**, 680–693 (2009)

Chapter 5
Heat Transfer Using the Local Thermal Non-Equilibrium Model

5.1 Input Parameters for the LTNE Model

For running the LTNE Model, the Reynolds number was also defined by Eq. (4.1). As mentioned before, for an impinging jet the flow is considered to be turbulent for $Re > 1,000$.

For simulating thermal non-equilibrium between phases, the low Re turbulence model presented in (Sect. 2.5) on pg. 13 was used is all simulations below. In order to guarantee that grids nodes be positioned within the laminar sub-layer, the closest grid node to the wall had a value for its wall coordinate y^+ less than unity ($y^+ \leq 1$). Further, inlet value for the turbulent kinetic energy k at the jet entrance was estimated using:

$$k_0 = \frac{3}{2}(v_0 I)^2 \tag{5.1}$$

where I is the turbulence intensity assumed to prevail in the incoming flow. For the dissipation rate of k, ε, the inlet value was calculated according to:

$$\varepsilon_0 = c_\mu^{3/4} \frac{k^{3/2}}{\ell} \tag{5.2}$$

where ℓ is a length scale associated with the energy containing eddies. Table 5.1 summarizes the parameters used as input.

5.2 Grid Independence Studies for the LTNE Model

For the LTNE model a set of testing calculation was performed for grid validation. The running condition were $Re = 10,400$, $H/B = 2.6$ and for an empty channel (Fig. 1.2a). At the jet entrance, values in Table 5.2 were employed. For grid

M. J. S. de Lemos, *Turbulent Impinging Jets into Porous Materials*,
SpringerBriefs in Computational Mechanics,
DOI: 10.1007/978-3-642-28276-8_5, © The Author(s) 2012

Table 5.1 Input parameters for turbulent flow simulations with the LTNE Model

Fluid	Density ρ	Viscosity μ	B	L	T_0	T_1	Length scale, l	Turbulence intensity, I
Air	1.225 kg/m^3	1.789×10^{-5} N.s/m^2	14.23 mm	500 mm	309.1 K	347.6 K	0.07B	2%

Table 5.2 Influence of grid size on integral wall heat flux—LTNE Model

Grid size	40×216	80×216	80×400
Wall heat power Q_w, Eq. (4.4)	772.75 W	818.87 W	825.68 W
Deviation in relation to grid 80×400	6.41%	0.82%	0.00%

independence studies, the overall heat power at the impinged wall was defined also considering the LTE hypothesis, giving,

$$
Q_w = \int\limits_{x=o}^{x=L} -q''|_{y=H}\, w\, dx, \quad q''|_{y=H}= \left(-k_f \frac{\partial \langle \bar{T} \rangle^i}{\partial y} \right)_{y=H}
\tag{5.3}
$$

where $w = 1$ m is the transverse plate width. Results for several grids are presented in Table 5.2. Also, as grid independency studies herein considered an empty channel only, local Nu number along that wall was evaluated with Eq. (2.51).

Figure 5.1a shows local Nu distribution calculated according to Eq. (2.51), also as a function of grid size. One can note in the table that for grids greater than 80×216, the deviation in relation to the finest grid is less than 1.0%. As such, all simulation for turbulent flow herein were carried out on a grid of size 80×216, which was refined close to the wall and about the jet entrance, where the steepest temperature gradients are expected to occur.

5.3 Empty Channel

Here, the first set of results is related to the configuration shown in Fig. 1.2a, where an empty channel is analyzed. Once an appropriate grid was chosen, code validation was carried out by comparing Nu numbers calculated at the bottom wall compared with results by Wang and Mujumdar [1], for two cases, namely for $H/B = 6$ and $Re = 5,200$ (Fig. 5.1b) and for $H/B = 2.6$ and $Re = 10,400$ (Fig. 5.1c). The figure indicates that for $H/B = 6$ a good agreement is obtained whereas for $H/B = 2.6$ results do not match quite well experimental values. This might be due to the fact that such flow entails a high degree of complexity, particularly for turbulent flow regime, as discussed by Wang and Mujumdar [1] and Heyerichs and Pollard [2]. Nevertheless, as the main purpose of this work is to investigate the influence of a porous layer on heat transfer, and not the turbulence model employed, and considering further the fact that a reasonable agreement with

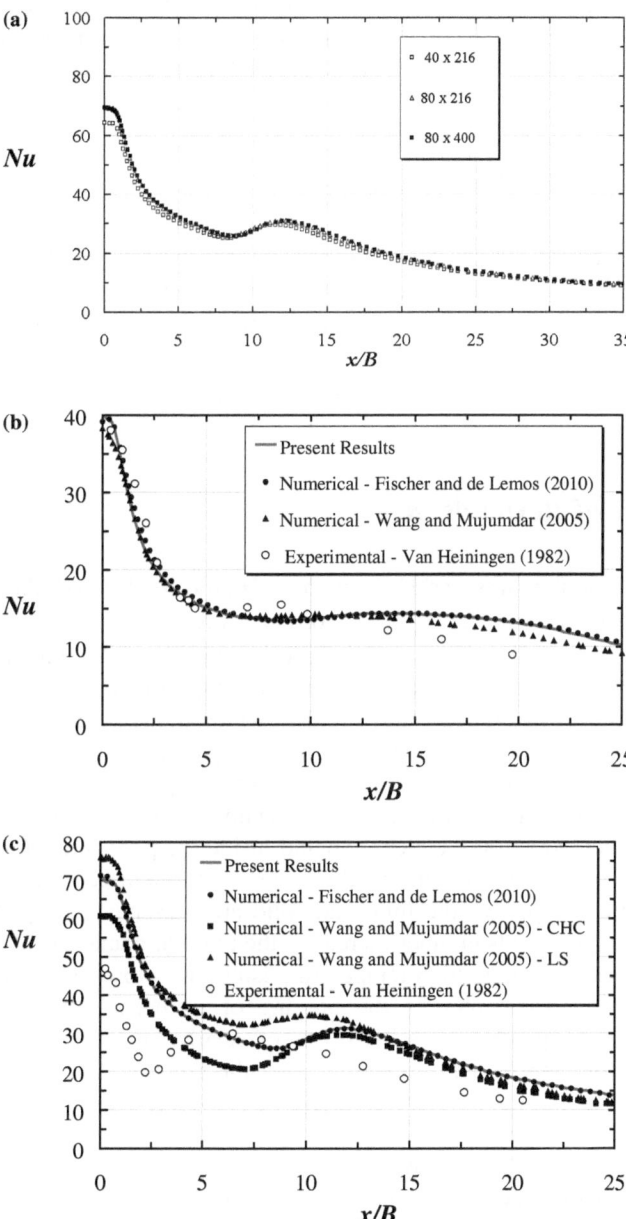

Fig. 5.1 Validation for distribution of *Nu* along the lower plate for clear channel. **a** Effect of grid size. **b** *Re* = 5,200, *H/B* = 6, exp. by Van Heiningen (1982) reported by [2]. **c** *Re* = 10,400, *H/B* = 2.6, Low *Re* models by Chang et al. (CHS) and Launder and Sharma (LS) reported by [1]

experimental data was achieved, the computer code and the grid size were assumed to be sufficiently accurate for the investigation here conducted. Results in Fig. 5.1 are also in agreement with those given by Fischer and de Lemos [3].

5.4 Channel with Porous Layer

When a layer of porous material is added to the bottom of the channel, the resulting configuration is shown in Fig. 1.2b. The material is assumed to be rigid, with porosity ϕ, non-dimensional thickness h/H, Darcy number $Da = K/H^2$ and thermal conductivity ratio k_s/k_f. Results below are obtained using various Re numbers and distinct values for such four parameters just mentioned.

5.4.1 Effect of Reynolds, Re

The Reynolds number was varied from $Re = 10,400$ to $Re = 60,000$ and results were obtained for $H/B = 2.6$, $k_s/k_f = 10$, $Da = 8.95 \times 10^{-5}$ and $h/H = 0.50$.

Figure 5.2a shows streamlines and indicates that as Reynolds number increases, the strength of the penetrating flow is sufficiently high to damp recirculation flow outside the porous material as more fluid flows though the permeable medium. Thin boundary layers occur not only around the jet exit but also at the stagnation region for high Re. Figure 5.2b shows corresponding results for the turbulent kinetic energy. One note that the higher the Reynolds number is, the higher are the turbulent kinetic energy on the clear gap and inside the porous medium as well, a feature that is in accordance with results by Graminho and de Lemos [4].

Temperature profiles for the fluid and solid are presented in Fig. 5.2c and d, respectively. As Reynolds number increases, the thermal boundary layer along the hot wall decreases, pushing high temperature isotherms close to the lower wall. It is also interesting to emphasize the bulging of isotherms at around $x/B = 7$ (x = 0.1 m), corresponding to a position of nearly stagnant flow. Past such position, the flow accelerates further reducing the thermal boundary layer thicknesses and enhancing temperature gradients at the lower wall.

Figure 5.3 shows corresponding effects of Re on Nu, calculated for both energy models. It is observed that for low Reynolds number, similar results with one and two energy equation model are obtained since for low Re the energy exchange between solid and fluid phase is lower when compared with high Reynolds number cases. As a consequence, the assumption of local thermal equilibrium hypothesis becomes more realistic for low Re flows where small convective currents promote less exchange of heat between phases. On the other hand, for high Reynolds number flows, the heat exchanged between phases becomes important and a substantially different local Nusselt is calculated depending on the model applied.

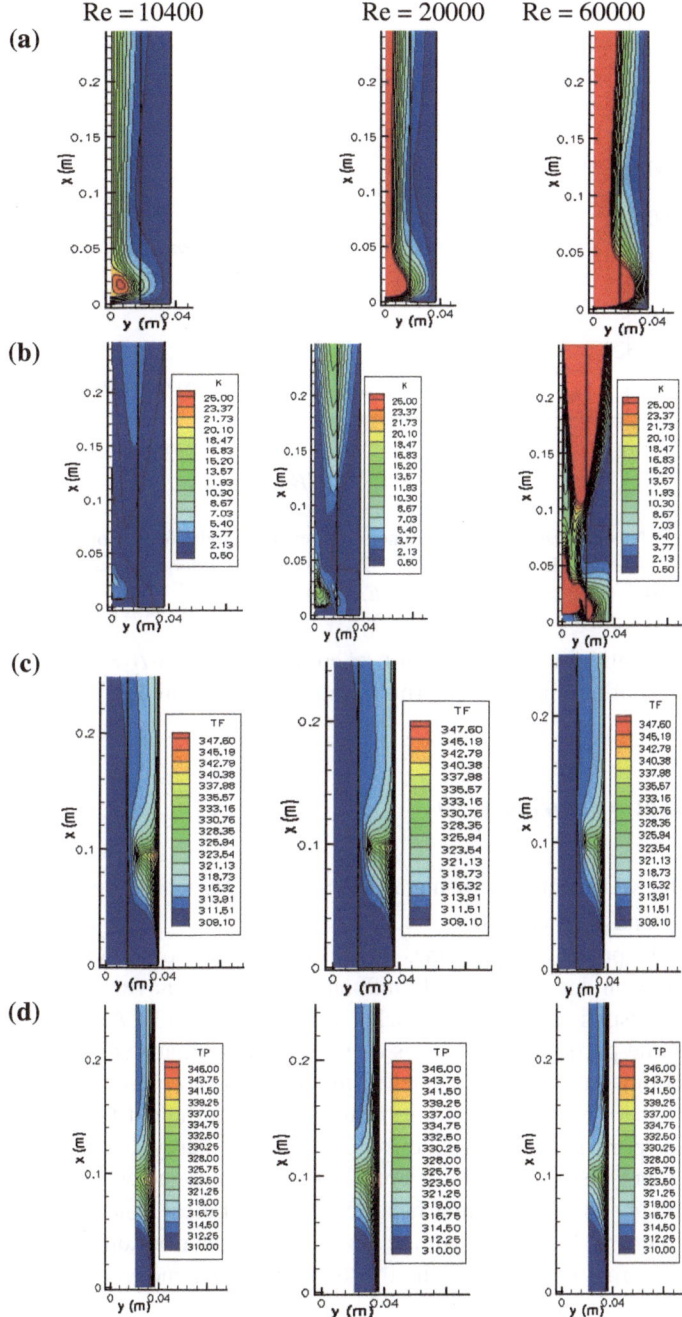

Fig. 5.2 Effect of *Re* for $H/B = 2.6$, $k_s/k_f = 10$, $Da = 8.95 \times 10^{-5}$, $h/H = 0.50$. **a** Streamlines. **b** Turbulent kinetic energy. **c** Fluid Temperature. **d** Solid Temperature

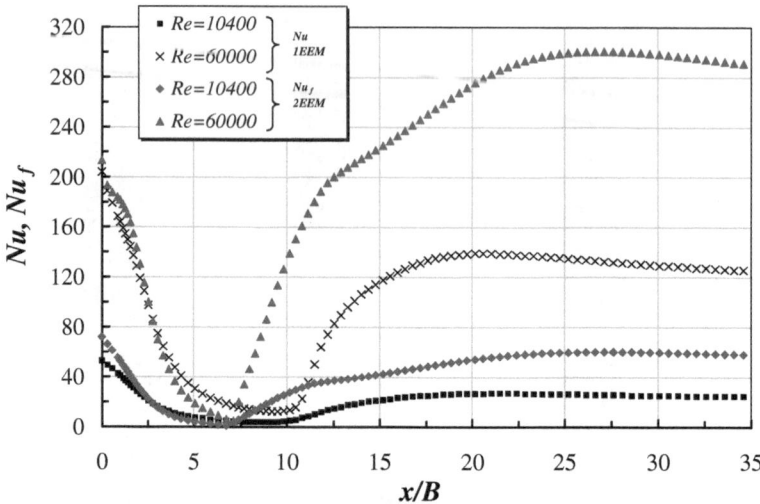

Fig. 5.3 Local Nusselt distribution as a function of energy model for various Reynolds number, 1EEM = One Energy Equation Model, 2EEM = Two Energy Equation Model, with $Da = 8.95 \times 10^{-5}$, $H/B = 2.6$, $k_s/k_f = 10$, $h/H = 0.50$

Also worth mentioning is the position of minimum Nu_f ($x/B = 7$), corresponding to the position of bulging of isotherms mentioned earlier in describing Fig. 5.3c, d.

5.4.2 Effect of Porosity, ϕ

Here, results were obtained using $H/B = 2.6$, $k_s/k_f = 10$, $Re = 10{,}400$, $Da = 8.95 \times 10^{-5}$ and $h/H = 0.50$.

Figure 5.4a presents streamlines for various values of ϕ. One can note that porosity variation does not strongly influence the flow behavior, as also confirmed by turbulent results presented in Fischer and de Lemos [3]. Also seen is that the presence of the porous layer reduces de size of the primary vortex when the flow pattern is compared to clear channel results (rightmost column in Fig. 5.4).

Figure 5.4b shows corresponding results for the turbulent kinetic energy. One note that as porosity decreases, turbulent kinetic energy levels increase. High values of k are also encountered around the jet entrance where steep velocity gradients occur. Around the interface, levels of turbulent kinetic energy are also high. This scenario contrasts with the clear channel distribution where most of the turbulence energy is generated in the recirculation zone corresponding to the primary vortex.

Results for fluid temperatures (Fig. 5.4c) seems to indicate that an increase in porosity decreases the thermal boundary layer thickness at the wall, affecting the temperature distribution within the liquid phase with a substantial cooling effect

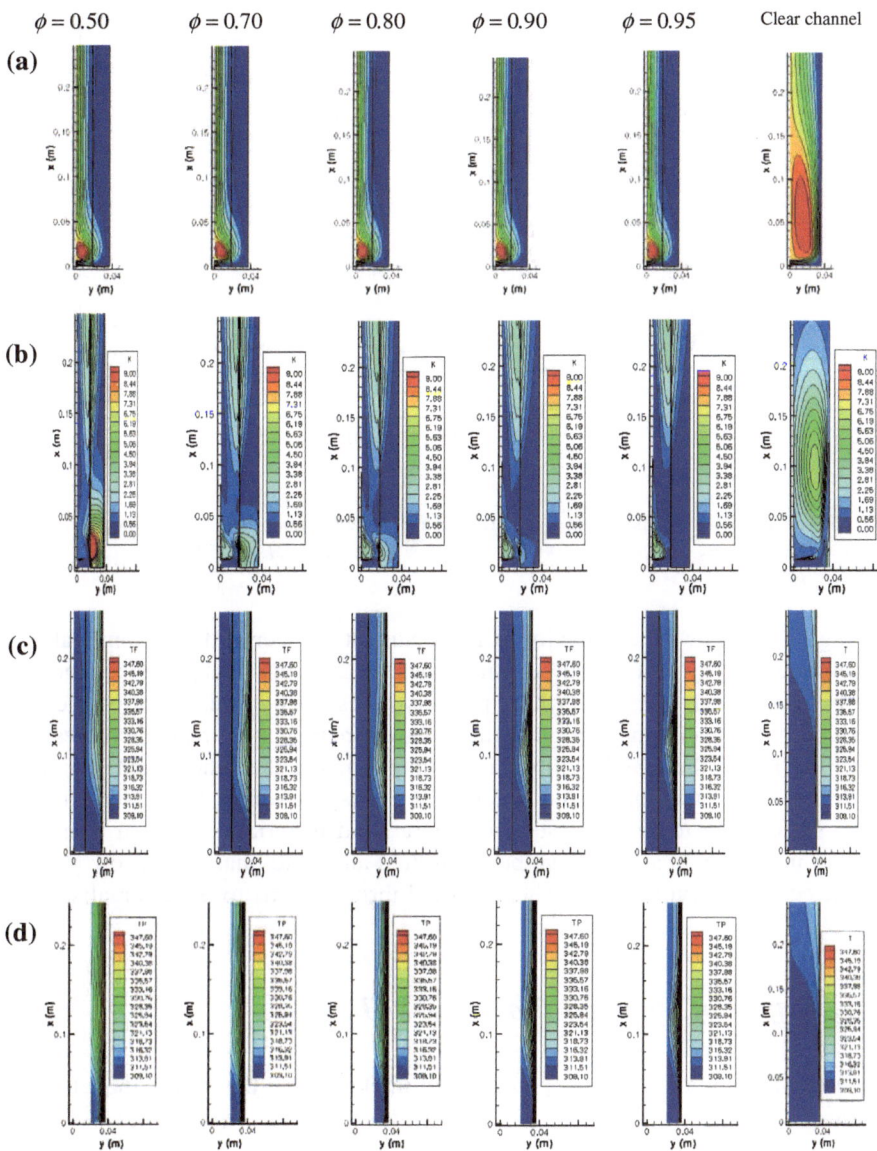

Fig. 5.4 Effect of porosity ϕ for $H/B = 2.6$, $k_s/k_f = 10$, $Re = 10,400$, $Da = 8.95 \times 10^{-5}$ and $h/H = 0.50$. **a** Streamlines. **b** Turbulent kinetic energy. **c** Fluid Temperature. **d** Solid Temperature

located at the stagnation region, a result also present in laminar simulations [5]. As φ increases, one can also note that as more void space is available for the fluid to permeate through the solid material, cooling of the porous substrate becomes more effective (Fig. 5.4d).

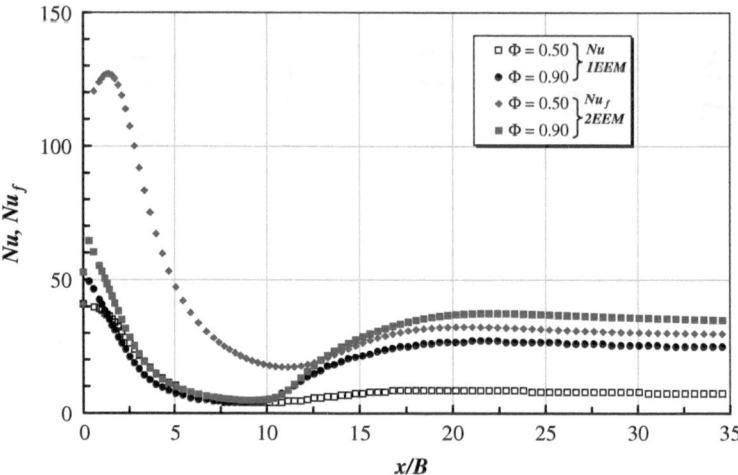

Fig. 5.5 Local Nusselt distribution as a function of energy model for various porosities with $H/B = 2.6$, $k_s/k_f = 10$, $Re = 10,400$, $Da = 8.95 \times 10^{-5}$ and $h/H = 0.50$

Figure 5.5 shows the behavior of local Nusselt number at the bottom wall as a function of energy model and porosity. Local Nusselt distribution calculated with the two models are similar for high porosity cases since in this situation the solid material influence is small. High φ values cases are then more realistic represented with the Local thermal Equilibrium assumption. On the other hand, for small porosities, there are substantial differences in distribution of local Nusselt numbers, since those represent cases with more solid material, which, in turn, have greater impact on heat exchange between fluid and solid phases. For low porosity cases, consideration of Local Thermal Equilibrium seems to be less realistic.

5.4.3 Effect of Channel Blockage, h/H

A study of the influence of the porous layer thickness and the energy model is now presented. Results are for $H/B = 2.6$, $k_s/k_f = 10$, $Re = 10,400$, $Da = 8.95 \times 10^{-5}$ and $\phi = 0.90$.

Figure 5.6a shows streamlines for several values of h/H. One can note that as the blockage ratio h/H increases, more fluid is forced through the porous material with accompanying reduction of the recirculating bubble in the gap region. The size of primary vortex is nearly extinguished for $h/H = 0.75$ and streamlines tends to become more uniform within the entire channel.

Turbulent kinetic energy behavior is presented in Fig. 5.6b where maps for k are shown. One can note that ratio h/H increases, turbulent kinetic energy is generated inside the porous material as well as in the free gap. Additional increase

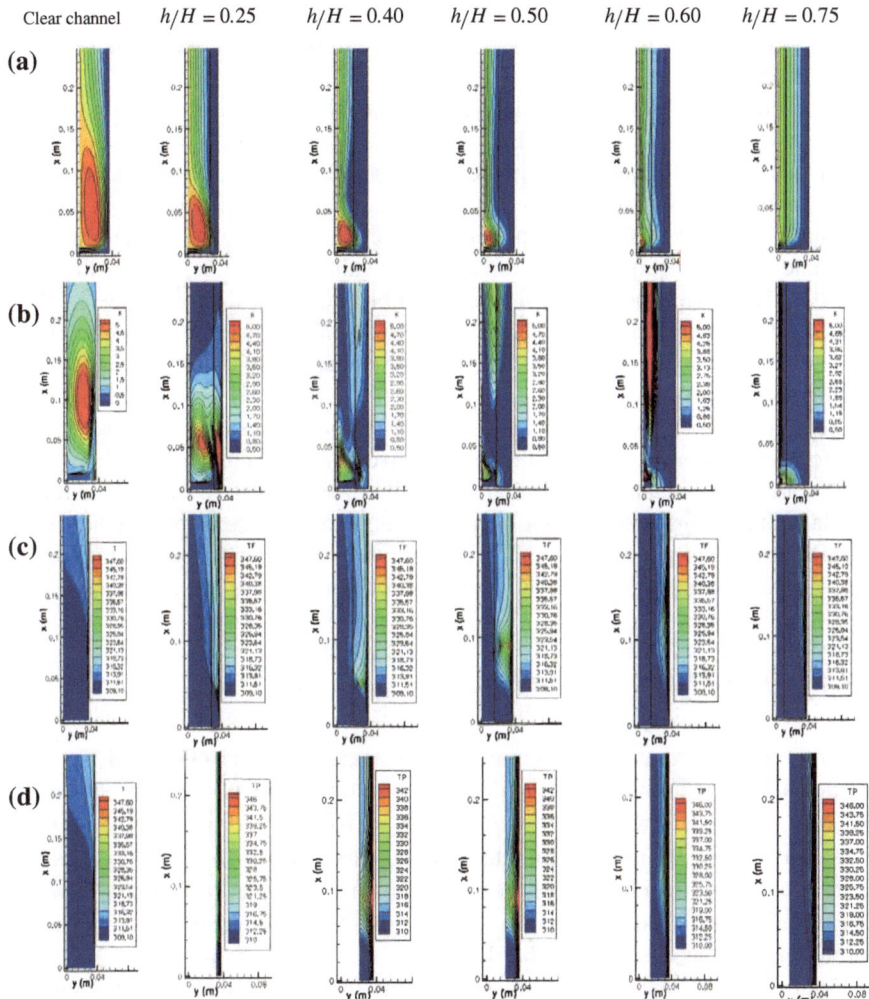

Fig. 5.6 Effect of blockage ratio h/H for $H/B = 2.6$, $k_s/k_f = 10$, $Re = 10{,}400$, $Da = 8.95 \times 10^{-5}$ and $\phi = 0,90$. **a** Streamlines. **b** Turbulent kinetic energy. **c** Fluid Temperature. **d** Solid Temperature

in the blockage ratio forces the flow to be aligned with the macroscopic interface, further downstream, leading to high generation rates of k around such interface. Further increase in h/H pushes the fluid outside the porous material, increasing the velocity gradient within the free gap, which, in turn, enhances k production rates therein. Effect of h/H on fluid temperatures are shown in Fig. 5.6c and show the enlarging of the thermal boundary layer as h/H increases, which present a peak that moves upstream and is reduced for larger blockage ratio. Corresponding results for T_s are presented in Fig. 5.6d, which indicates that for low blockage ratios solid and fluid temperatures are quite different, but tend towards equality as ticker porous layers allow for more room for equilibrium between phases to be established.

As in previous plots for *Nu*, Fig. 5.7a presents Nusselt numbers calculated with the Local Equilibrium Model (1EEM), Eq. (2.51), with those obtained with the Local Non-thermal Equilibriun model (2EMM), Eq. (2.52). For thinner porous layers, both models present a second peak on *Nu* with and values for it that are substantially different for each model, a reflect of the fact that temperatures for solid and fluid are quite distinct for low h/H values (Fig. 5.6c, d). On the other hand, for thicker porous layers models exhibit similar behaviors since in these cases heat exchange between phases is more complete and, as a consequence, temperatures in both phases tend be equal. Under those conditions, the Local thermal Equilibrium assumption turns to be more realistic. This observation can be further noted Fig. 5.7b, which was plotted for the position $x/B = 2.5$ and confirms that for thicker porous layers differences between temperatures of solid and fluid phases are small, leading to similar values for the respective Nusselt numbers. Similar conclusions were drawn by [5] for laminar flow, except that here temperature gradients are higher than in laminar flow since in turbulent regime the inter-phase heat exchange, as well as wall heat transfer, are more intense.

5.4.4 Effect of Darcy Number, Da

Differently from what was observed in the analysis of the effects of porosity, permeability does influence the hydrodynamic field as sown in Fig. 5.8a. Results are here calculated with $H/B = 2.6$, $k_s/k_f = 10$, $Re = 10,400$, $\phi = 0.50$ and $h/H = 0.50$. The primary vortex intensity decreases with a decrease in *Da* as well as the fluid speed inside the porous layer, which leads to a lower mass flow rate inside the porous substrate. For $Da \leq 2.89 \times 10^{-6}$ the flow nearly vanishes in the porous material as can be seen by the streamlines in Fig. 5.8a.

Figure 5.8b presents two-dimensional fields for the turbulence kinetic energy, *k*. For $Da = 7.30 \times 10^{-3}$ and $Da = 3.68 \times 10^{-4}$, generation of *k* is enhanced and is even grater than for the clear channel case. The region of highest turbulence corresponds to the jet entrance where steep velocity gradients produce high generation rates of *k*. For very low permeabilities, $Da = 4.86 \times 10^{-7}$, as fluid is pushed to flow thought the clear gap, higher values of *k* are found around the macroscopic interface between the two media. In addition, high *k* values are also found due to the small recirculation bubble attached to the jet entrance.

Further, Fig. 5.8c, shows temperature distributions for distinct values of $Da = K/H^2$. One can see that as permeability decreases, isothermal lines, referent to the highest temperatures, bulge from the hot wall thickening the thermal boundary layer at the wall surface. Cases with low permeability present fluid temperature peaks at around $x = 0.06$m for $Da = 2.89 \times 10^{-6}$ and $x = 0.04$m for $Da = 4.86 \times 10^{-7}$. Those peaks seems to be related to flow blockage, which

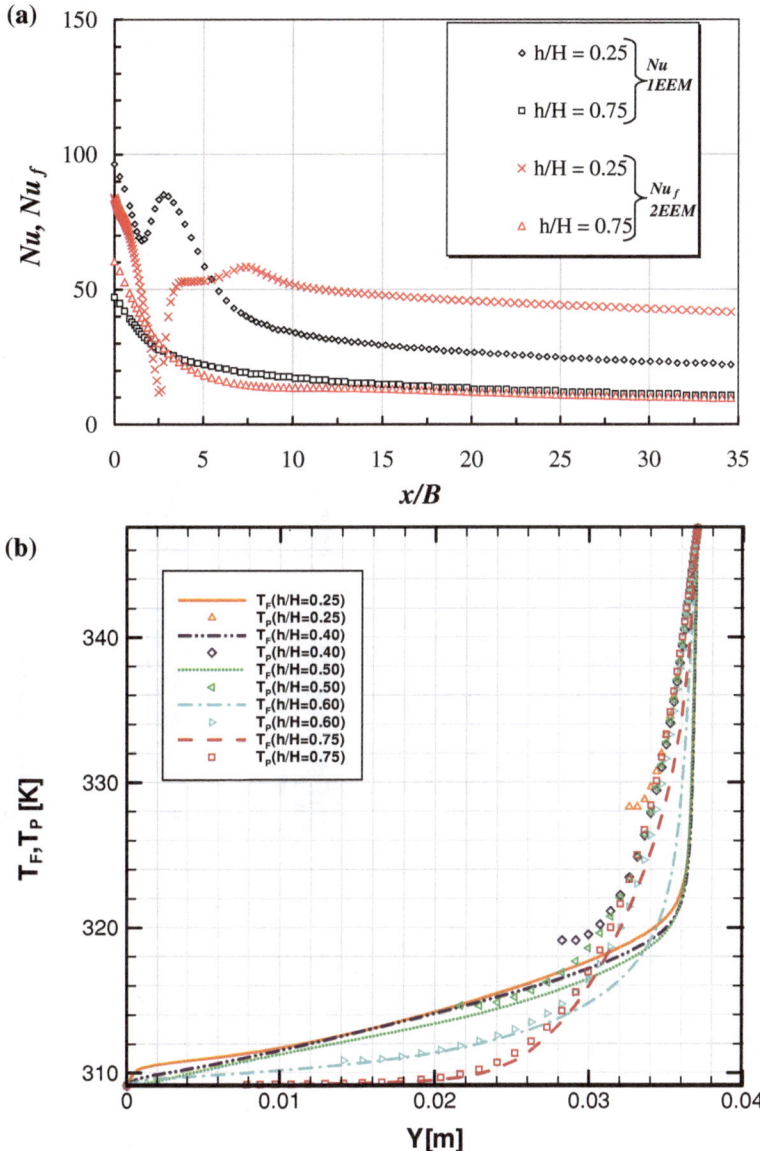

Fig. 5.7 Comparison of axial Nusselt distribution (**a**) and transversal temperature profiles at $x/B = 2.5$ for various blockage ratios h/H with $H/B = 2.6$, $k_s/k_f = 10$, $Re = 10,400$, $Da = 8.95 \times 10^{-5}$ and $\phi = 0.90$

happens earlier along the flow (smaller x/B) as Da is reduced. Corresponding T_s results are found in Fig. 5.8d where corresponding peaks in the solid temperature can be identified for low permeabilites ($Da \leq 2.89 \times 10^{-6}$).

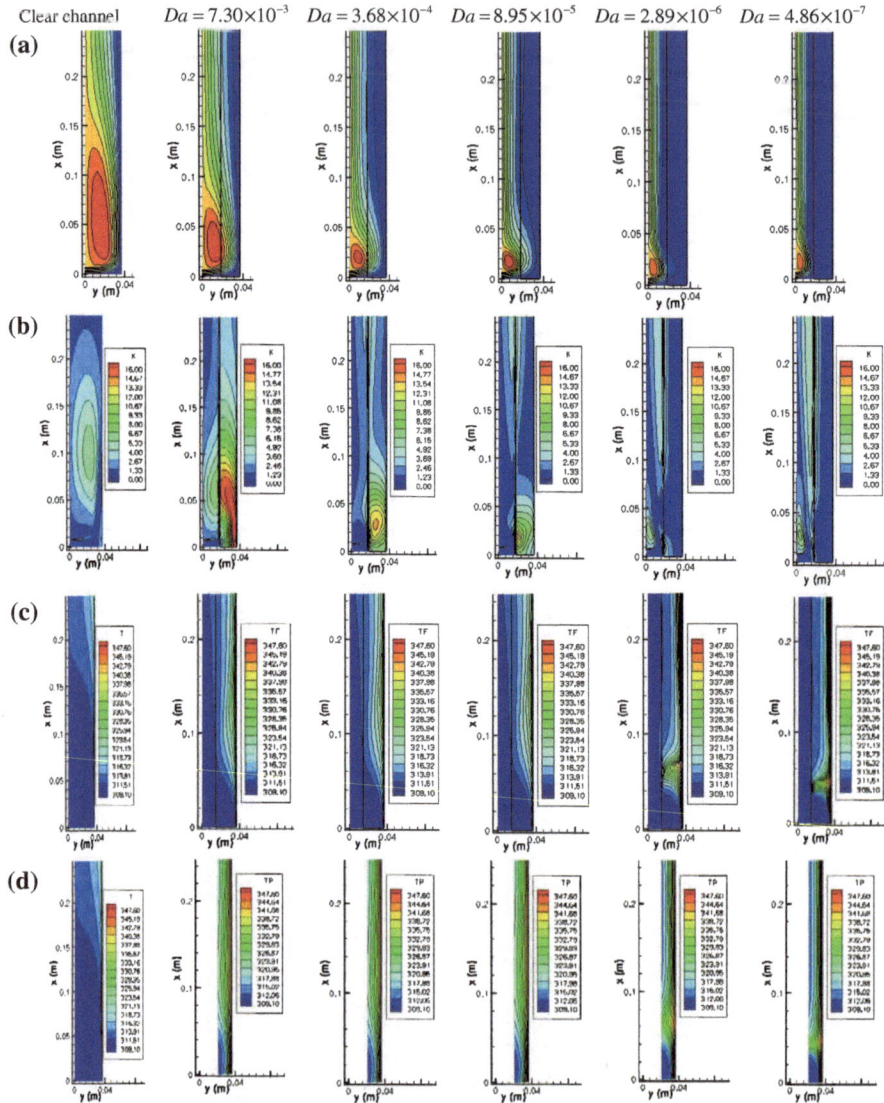

Fig. 5.8 Effect of Da for $H/B = 2.6$, $k_s/k_f = 10$, $Re = 10{,}400$, $\phi = 0.50$ e $h/H = 0.50$.
a Streamlines. **b** Turbulent kinetic energy. **c** Fluid Temperature. **d** Solid Temperature

Figure 5.9 shows Nusselt numbers along the wall as a function of Da, where one can see that in such cases Nu at the stagnation region is most influenced, being reduced as Da is reduced. In the wall jet region, local Nu is also reduced, but less intensively when compared with its variation as a function of porosity (Fig. 5.4). Also important to note is that for low Da, Nusselt values using either 1EEM or

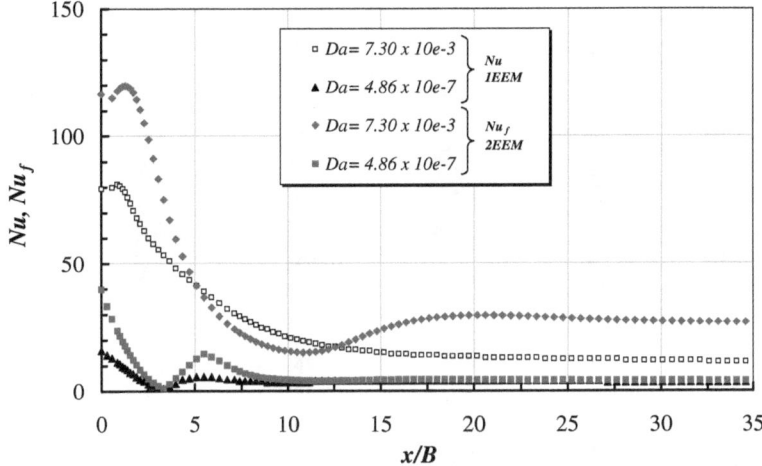

Fig. 5.9 Local Nusselt distribution as a function of energy model for Darcy number with $H/B = 2.6$, $k_s/k_f = 10$, $Re = 10{,}400$, $h/H = 0.50$ and $\phi = 0.50$

2EEM are nearly the same in the developed region ($x/B > 20$), because for very low mass flow rates inside the porous bed, both the solid and fluid temperature gradients at wall attain nearly the same value, giving further the same Nu number and validating the so-called thermal equilibrium assumption within the porous material.

5.4.5 Effect of Solid-to-Fluid Thermal Conductivity Ratio k_s/k_f

The effect of k_s/k_f on flow and thermal fields is presented in Fig. 5.10, based on the parameters $H/B = 2.6$, $Re = 10{,}400$, $Da = 8.95 \times 10^{-5}$, $\phi = 0.90$ and $h/H = 0.50$.

As expected, the mean (Fig. 5.10a) and statistical (Fig. 5.10b) flow structures are not influenced by the solid-to-fluid thermal conductivity ratio since in the solution methodology applied decoupled formulation was used. Or say, the calculated flow field was not impacted by the thermal field since only constant property cases were run.

Fluid temperatures are presented in Fig. 5.10c and indicate the thermal field is significantly affected by the value of k_s/k_f, as expected. Increasing k_s/k_f reduces the fluid temperature gradient at the wall, as a consequence of reducing the solid temperate at corresponding locations (Fig. 5.10d). High k_s/k_f ratios increase solid temperature elsewhere, bringing up fluid temperatures as a result of interfacial heat transfer between phases. Also interesting to note is the result on the thermal filed of the nearly stagnant flow around $x/B = 6.5$, where both the fluid and the solid

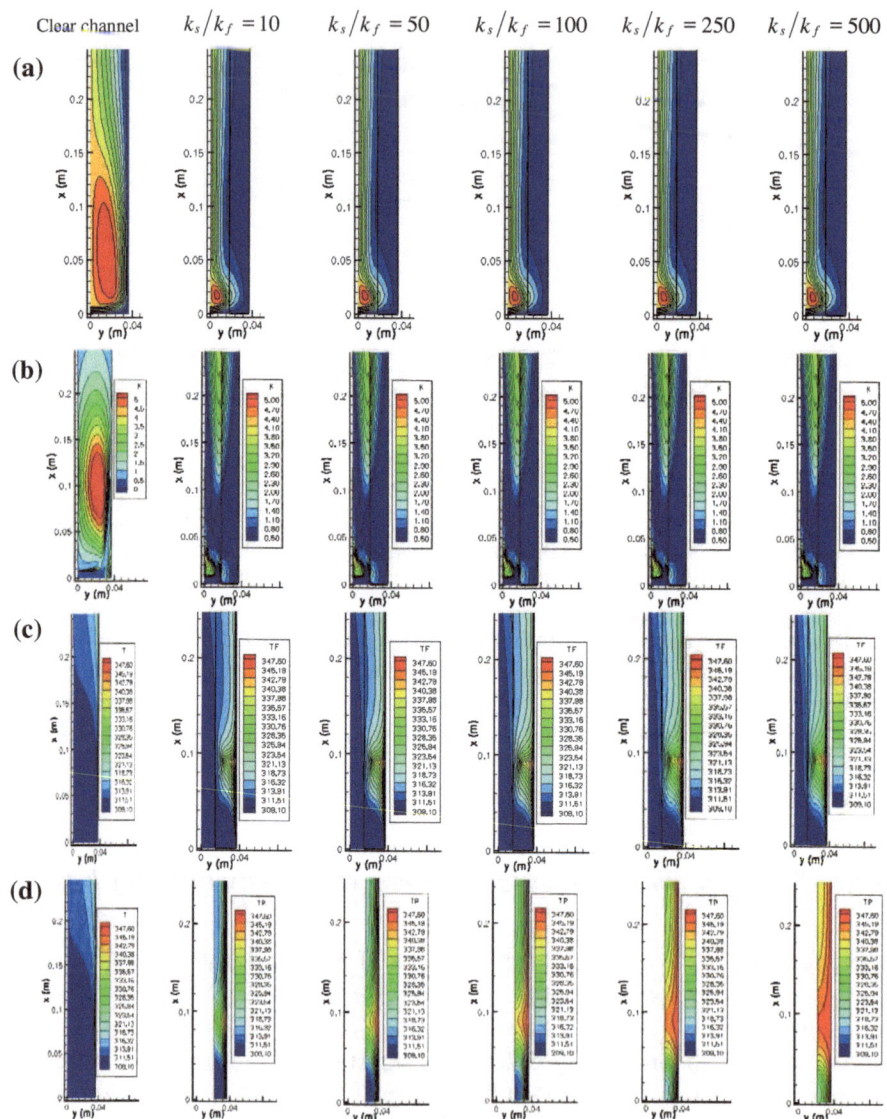

Fig. 5.10 Effect of k_s/k_f for $H/B = 2.6$, $Re = 10,400$, $Da = 8.95 \times 10^{-5}$, $\phi = 0.50$ e $h/H = 0.50$. **a** Streamlines. **b** Turbulent kinetic energy. **c** Fluid Temperature. **d** Solid Temperature

isotherms depart from the wall, with consequent thickening of thermal boundary layers and reductions of temperature gradients at the wall. Nusselt number results in Fig. 5.11 corroborates such findings. One can see in the figure that for the 1EEM model there is a substantial reduction of Nu as k_s/k_f increases, and that when using

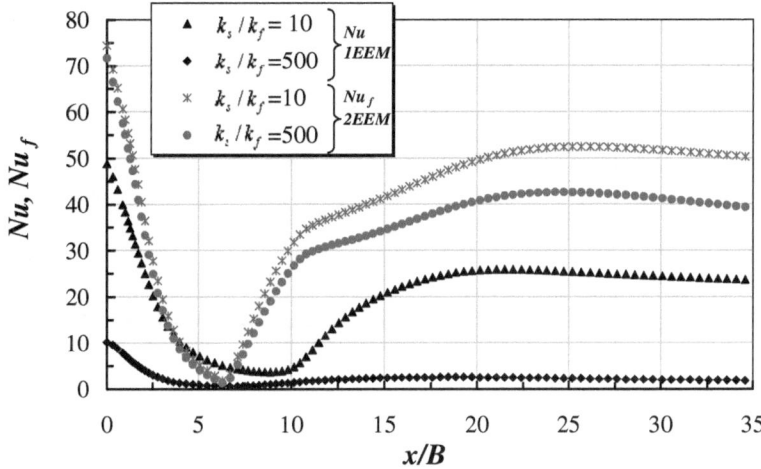

Fig. 5.11 Local Nusselt distribution as a function of energy model for various ratios k_s/k_f with $H/B = 2.6$, $Re = 10{,}400$, $Da = 8.95 \times 10^{-5}$, $\phi = 0.90$ and $h/H = 0.50$

the local non-thermal equilibrium hypotheses (2EEM) no such substantial reduction in Nu_f is calculated. Also interesting to see is the reduction on Nu_f at about $x/B = 6.5$, which reflects the bulging of isotherms seen earlier (Fig. 5.10c, d).

5.5 Integral Wall Heat Flux

In an earlier paper [4], is was observed that the overall heat transfer to or from a wall would be an indicator of the effectiveness in using porous layers attached to surfaces. Such overall heat transferred from the lower wall to the flowing fluid can be calculated for both configurations presented in Fig. 1.2, as

$$q_w = \frac{1}{L}\int_o^L q_{wx}(x)\, dx \qquad (5.4)$$

Depending on model used, there are two possibilities to evaluate the local wall heat flux q_{wx}. One can use the hypothesis of Local Thermal Equilibrium (LTE or 1EEM), or else, individual terms can be applied in each phase in order to calculate the integrated heat transferred from the bottom wall. In the latter case, the LTNE (2EEM) model is employed.

Therefore, for the one-energy equation model, one has:

$$q_w = \frac{1}{L}\int_o^L q_{wx}(x)\, dx; \quad q_{wx} = -k_{eff}\frac{\partial \langle T \rangle^i}{\partial y}\Bigg|_{y=H} ; \quad k_{eff} = \phi k_f + (1-\phi)k_s \qquad (5.5)$$

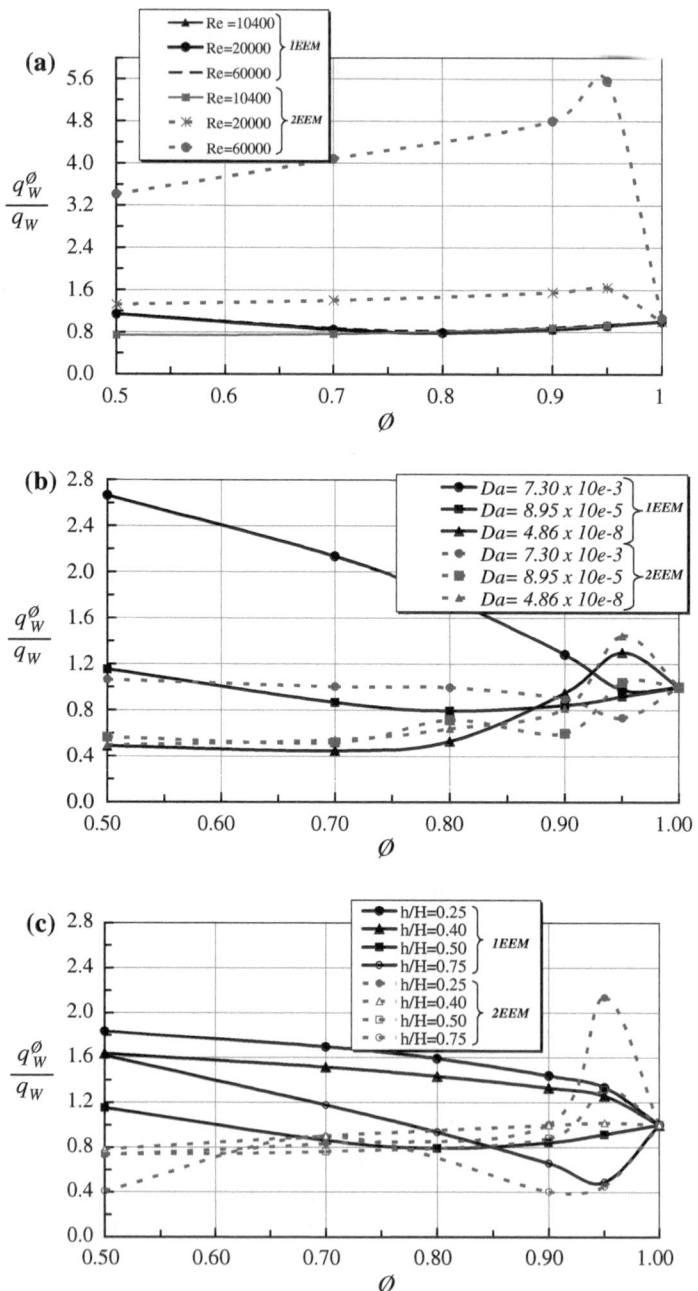

Fig. 5.12 Integral heat flux ratio at the lower wall for varying porosity ϕ, $H/B = 2.6$ and $k_s/k_f = 10$. **a** Effect of Re. **b** Effect of Da. **c** Effect of h/H

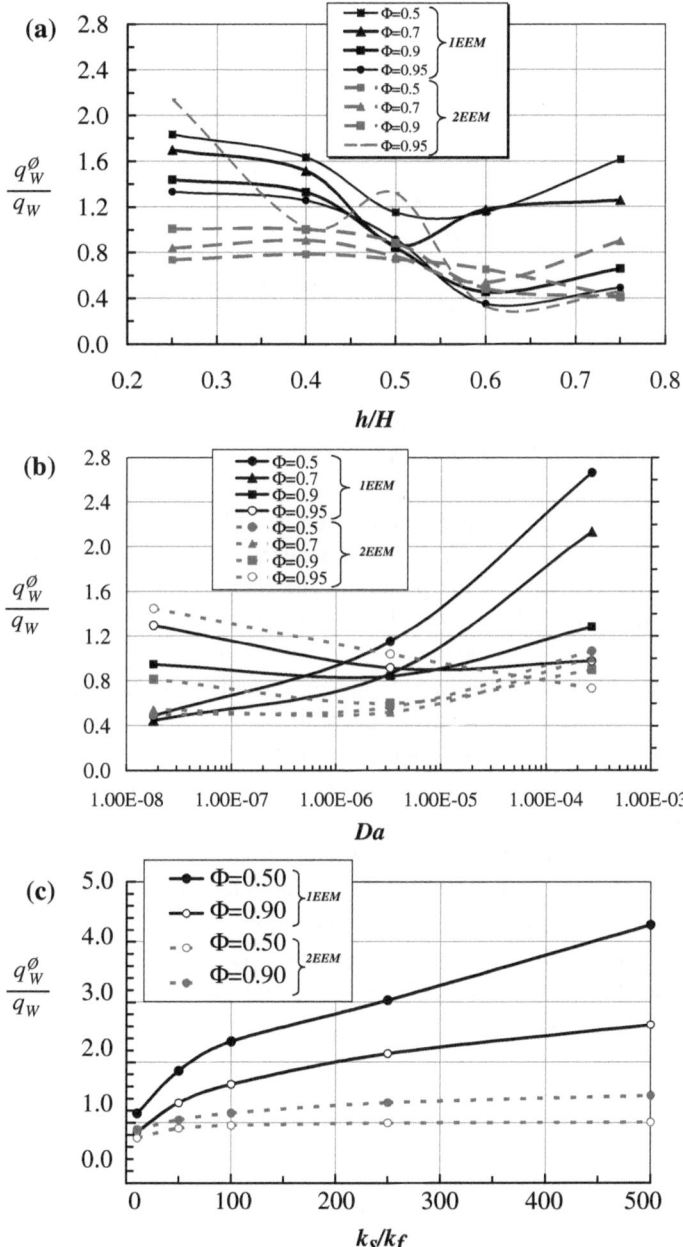

Fig. 5.13 Integral heat flux ratio at the lower wall for distinct porosities ϕ with $H/B = 2.6$, $Re = 10,400$. **a** Varying h/H, $Da = 8.95 \times 10^{-5}$, $k_s/k_f = 10$. **b** Varying Da, $h/H = 0.5$, $k_s/k_f = 10$. **c** Varying k_s/k_f, $Da = 8.95 \times 10^{-5}$, $h/H = 0.5$

and for the two-energy equation model:

$$q_w = \frac{1}{L} \int_o^L q_{wx}(x)\, dx;$$

$$q_{wx} = - \left[k_{eff,f} \frac{\partial \langle T_f \rangle^i}{\partial y} \bigg|_{y=H} + k_{eff,s} \frac{\partial \langle T_s \rangle^i}{\partial y} \bigg|_{y=H} \right]; \quad \begin{cases} k_{eff,f} = \phi k_f \\ k_{eff,s} = (1-\phi)k_s \end{cases}$$

$$(5.6)$$

For the cases where the a porous layer is considered, the wall hat flux is given a superscript ϕ on the form q_w^ϕ. The ratio q_w^ϕ / q_w can then be seen as a measure of the effectiveness of using a porous layer for enhancing or damping the amount of heat transferred through the wall. In the results for q_w^ϕ / q_w to follow, *solid lines* are a representation of simulation with the LTE model (1EEM) whereas LTNE computations are plotted using *dotted lines* (2EEM).

Figure 5.12 compares the ratio q_w^ϕ / q_w as function of ϕ for distinct Re, Da and h/H, with $k_s/k_f = 10$. For high Reynolds number, Fig. 5.12a indicates that using a porous layer is always beneficial to heat transfer when the 2EEM model is applied, regardless of the porosity of the medium. The figure also suggests that for cases with high permeabilities (high Da), there is also a net gain when using simulations are computed with either model ($q_w^\phi / q_w > 1$). Figure 5.12c further suggests that when using the 2EEM model, only highly permeable (high ϕ) and thinner layers (small h/H) favor heat transfer from the wall.

Figure 5.13 further indicates that for highly porous (Fig. 5.13a) and highly permeable materials (Fig. 5.13b), results for q_w^ϕ / q_w under the LTNE assumption (*dotted lines*) give values greater than unity, a result that generally contrasts with those by [3] who use the LTE approach under the same conditions (*solid lines*).

Finally, Fig. 5.13c compiles results for q_w^ϕ / q_w when the conductivity ratio k_s/k_f is varied. Here, one can notice that only for high porosities and for $k_s/k_f > 10$, both models suggest that the use of a porous layer can always enhance the heat extracted from the bottom wall. The results herein might be useful to design and analysis of energy efficient equipment.

References

1. S.J. Wang, A.S. Mujumdar, A comparative study of five low Reynolds number $k - \varepsilon$ models for impingement heat transfer. Appl. Therm. Eng. 25, 31–44 (2005)
2. K. Heyerichs, A. Pollard, Heat transfer in separated and impinging turbulent flows. Inter. J. of Heat and Mass Transfer **39**(12), 2385–2400 (1996)
3. C. Fischer, M.J.S. de Lemos, A turbulent impinging jet on a plate covered with a porous layer. Numer. Heat Transf. **58**, 429–456 (2010)

4. D.R. Graminho, M.J.S. de Lemos, Simulation of turbulent impinging jet into a cylindrical chamber with and without a porous layer at the bottom. Inter. J. Heat Mass Transf. **52**, 680–693 (2009)
5. F.T. Dórea, M.J.S. de Lemos, Simulation of laminar impinging jet on a porous medium with a thermal non-equilibrium model, Inter. J. Heat and Mass Transf. **53** 5089–5101 (2010)

Chapter 6
Concluding Remarks and Future Work

This book investigated the influence of the presence of a porous layer covering a surface where a jet collides. This work reviewed and compiled a systematic study on impinging jets on bare and covered walls, which was carried out in the last few years at ITA, Brazil, and considered both laminar [1–3] and turbulent flow regimes [4–6]. By that, a self-contained text was put together in order to convey to the interested reader the major steps and results achieved on such research topic.

Two energy modes were applied, namely 1EEM and 2EEM, based respectively on the Local Thermal Equilibrium (LTE) and Local Thermal Non-Equilibrium hypotheses (LTNE). It was observed that the Reynolds number and porosity strongly influences the stagnation Nusselt value while the porous layer thickness affects more intensely the distribution of Nu along the plate. Cases with low porosity and highly permeable layers of porous material tend to yield better heat absorption/release rates when compared with a bare wall case. Regardless of the model used, increasing the thermal conductivity ratio is always beneficial to heat transfer enhancement form the hot wall. Ultimately, results in this work might be useful to engineers designing systems that make use of impinging jets over thermally conducting porous materials.

Future work may take into consideration chemical reactions within the fluid phase and the movement of the solid phase. By that, a more complete and more general model would be available contributing to solutions of a broader range of problems, in a more realistic fashion, including simulation of modern equipment for gasification of renewable fuels and for advanced materials production.

References

1. D.R. Graminho, M.J.S. de Lemos, Laminar confined impinging jet into a porous layer. Numer. Heat Transf. Part A Appl. **54**(2), 151–177 (2008)
2. M.J.S. de Lemos, C. Fischer, Thermal analysis of an impinging jet on a plate with and without a porous layer. Numer. Heat Transf. Part A Appl. **54**, 1022–1041 (2008)

M. J. S. de Lemos, *Turbulent Impinging Jets into Porous Materials*,
SpringerBriefs in Computational Mechanics,
DOI: 10.1007/978-3-642-28276-8_6, © The Author(s) 2012

3. F.T. Dórea, M.J.S. de Lemos, Simulation of laminar impinging jet on a porous medium with a thermal non-equilibrium model. Int. J. Heat Mass Transf. **53**, 5089–5101 (2010)
4. D.R. Graminho, M.J.S. de Lemos, Simulation of turbulent impinging jet into a cylindrical chamber with and without a porous layer at the bottom. Int. J. Heat Mass Transf. **52**, 680–693 (2009)
5. C. Fischer, M.J.S. de Lemos, A turbulent impinging jet on a plate covered with a porous layer. Numer. Heat Transf. Part A Appl. **58**, 429–456 (2010)
6. M.J.S. de Lemos, F.T. Dórea, Simulation of turbulent impinging jet into a layer of porous material using a two-energy equation model. Numer. Heat Transf. Part A Appl. **59**(10), 769–798 (2011)